The Internet as a Diverse Community

Cultural, Organizational, and Political Issues

Urs E. Gattiker
Aalborg University, Denmark

LEA LAWRENCE ERLBAUM ASSOCIATES, PUBLISHERS

2001   Mahwah, New Jersey   London

Lawrence Erlbaum Associates, Inc., Publishers
10 Industrial Avenue
Mahwah, NJ 07430

Cover design by Jennifer Anne Sterling

Library of Congress Cataloging-in-Publication Data

Gattiker, Urs E.
The Internet as a diverse community : cultural, organizational, and political issues / by Urs E. Gattiker.
p. cm.
Includes bibliographical references and index.
ISBN 0-8058-2488-X (cloth : alk. paper) — ISBN
0-8058-2489-8 (pbk. : alk. paper) 2/ 289255
1. Internet—Social aspects.
2. Telematics—Social aspects. 3. Pluralism (Social sciences). 4. Community. I. Title.
HM851 .G37 2000
303.48'33—dc21

99-054567
CIP

Books published by Lawrence Erlbaum Associates are
printed on acid-free paper, and their bindings are chosen for
strength and durability.

Printed in the United States of America
10 9 8 7 6 5 4 3 2 1

To Inger Marie and Melanie
who continue to enrich my life

Contents

Read / Paraphrase

Something for Everyone:
A Reader's Guide

This section is for those who may not have the time to read a book from beginning to end. As a part of the text, I present a list of abbreviations and extensive indices (author, subject, geographical location, and firms/associations mentioned) to assist readers. This enables readers to easily locate specific topics long after having finished reading this book. For the Internet novice, or for those interested in more detailed information on technical matters, appendices are provided.

One should begin this book at its beginning and end at its end; yet, en route, some sections and subsections may be omitted without losing the grasp of the main argument. Chapter overviews are provided to give the reader a quick review of the topics and issues addressed and to map out the ideas and theories discussed. The reader may then discriminate, according to his or her individual needs. Appendices listed under each overview provide additional information relevant to the content. To illustrate models, thoughts, and frameworks, practical examples and cases are used from organizations and countries from around the world. (e.g., Australia, Europe, Africa, and Asia). Such examples epitomize the ease with which the theories and methods presented herein can be applied to genuine scenarios.

Part I sets the stage for the reader. Chapter 1 provides a basic overview of the topic and introduces the reader to computer-mediated communication and the Internet. Chapter 2 focuses on regulatory developments in the telecommunications sector and how these have influenced the use of the Internet by both business and private parties. Chapter 3 outlines how issues addressed in chapters 1 and 2 affect the way in which people and organizations gain access and the cost of surfing on the Web as incurred by business and private users.

Part II focuses on cultural, attitudinal, and other influences on people's use of the Internet. Regulatory issues as outlined in part I either reduce or exacerbate differences between user groups in various countries due to cultural characteristics as discussed in chapter 4. Chapter 5 shows how issues outlined earlier, and especially regulatory and cultural factors may help explain ethical and moral issues as experienced by users to increase safety and security of information exchanged on the Internet while reducing the risks for disasters such as viruses and hacking. Chapter 6 further outlines how the internationalization of business, with the help of the Internet, requires some codes of conduct for users, clients, firms, and associations to reduce misunderstandings and conflict. The discussions show that we are still far away from sharing ethical and moral standards, thus making it difficult to develop rules and norms for people around the world.

Part III of the book outlines some of the issues that still must be addressed by researchers and managers as well as by policymakers alike to increase the peaceful and beneficial use of the Internet for the advantage of many stakeholders with divergent interests. Chapter 7 outlines some of the research questions that still await investigation to further explain why people may do certain things on the Internet (e.g., downloading of pornographic files) but refrain from doing others (e.g., shopping for groceries on the Web). Chapter 8 provides a summary of the book and conclusions as well as implications for decision makers and researchers. Moreover, the difficulty with developing virtual communities and institutional characteristics for the Internet is outlined and linked to earlier chapters in the book.

The appendices provide additional information and insights (e.g., glossary of terms, code of conduct for privacy and safety with data and information, how to do information searches online and how to cite resources found on the Internet, cost control, caller ID information and tidbits).

This book fails to touch on many topics. Contemporary concerns and questions, such as how technical innovation really takes place, are left alone. The discussion of cases, research and development (R&D), or technology management in general (e.g., Gattiker, 1990c) is kept to a minimum. This book is not meant to go into great depths about technology-related issues in the more classical sense and how they affect the Internet. Technology is not just limited to R&D and the discussion of the latest technology gadget. As history has taught, it can take years before we discover a specific technology's most beneficial use (i.e., the steam engine or laser technology). Instead, the book focuses on social systems, particularly governments, organizations, cities and countries, and consumers and society as a whole. The assumption is that individual behavior, in some aggregate form, drives the activity of a social system. As part of the electorate, an individual may influence the parameters that affect the technology's penetration of society.

This book does not attempt to provide readers with ready solutions; instead, it aims to provide a framework that will promote discussion of pertinent issues. Instead of addressing narrow topics with considerable detail, I hope to add breadth to the dialogue between the advocates and adversaries of the Internet and multimedia, while maintaining as much depth as possible. An attempt to reach a critical synthesis of the huge body of literature that includes communication, economics, philosophy, political science, management, psychology, science policy, sociology, and telecommunication engineering is one of the goals of this text. Furthermore, managerial elements, social policy, and ethical and moral arguments are used to illustrate the various applications of Internet-related technology as a continuum, rather than as a dichotomy.

As an emerging institution, the Internet does have some characteristics of well-established institutions we all know about (e.g., most churches). However, many features are unique, such as the limited importance of geographical boundaries and how this exposes us to people with vastly different values, norms, and beliefs as reflected by their varied cultural backgrounds and interests. Hence, the Internet is in part a vast laboratory where new ideas, social groups, interests, and concerns are discussed between individuals who may never have met in person and who may never will. How such an institution is evolving and where its positive and less positive outcomes can be seen, as well as what the future might hold, is at the core of this book.

A great deal of research has been done with little or no return in the development of our understanding; instead, knowledge for its own sake has been the result. This book intends to provide a treatise of the more important aspects of Internet-related matters. It is vital that students of the Internet and cyberspace-related technologies understand how ethics, morals, communication, marketing, justice, and individual rights and responsibilities affect the global competitiveness of firms and nations using the Internet. Most importantly, the link between these issues and the possible creation of an institution without apparent geographical or legal boundaries (maybe fluid ones) is a timely topic that is examined. The uninformed and imprudent use of the Internet is frequently unjust and may infringe upon the moral rights of other individuals and nations.

This book gives the general reader and the student of Internet issues a survey that points out accomplishments, issues, and problems. The primary

objective of this book is to provide a broad view of Internet/technology-re-
lated theory; it is not a review of current literature. It is my aim to provide a
treatise that integrates the ideas and thoughts of the various disciplines that
address and discuss Internet matters.

How Can This Book Be Used in My Studies or To Help Solve Important Issues and Make the Best Decisions Possible?

Most principles and methods are outlined in tables and figures. In turn, because the reader is continuously referred to tables and other chapters for
further explanations, one can easily refresh his or her understanding of the
most pertinent issues. The indices also help save time for the individual
who needs to quickly locate a section of the book. In addition to the extensive subject index, the book also offers a geographical location index.

For the person interested in business and political issues, I have provided
an index dedicated specifically to companies and associations. Hence, one
can easily find examples as they pertain to one firm and/or industry
throughout the book. Additionally, focusing on how concerns are handled
by various associations, as well as by various interest and lobby groups, is
facilitated with this index.

How Does This Book Anticipate and Dispense With the Further Globalization of Internet Issues?

This volume addresses Internet and multimedia issues from an international perspective by outlining issues of international sovereignty and the
potential impact of national interests on global technology policy. Some
important concerns are illustrated by examples from geographically dis-
persed areas.

As is to be expected of a Canadian author, the focus rests mainly on the
issues of Canada, the United States, and Mexico in response to the growing
importance of trade between these countries. Similarly, dual Swiss–Canadian citizenship allows me to add a Swiss perspective to these issues. Furthermore, my work in Australia, South Africa, Germany, and Denmark
allows me to discuss the issues and concerns raised in these countries.

Regardless of the attempts to acknowledge the globalization of technology-related issues, trade, and policies, this text is heavily influenced by the
thoughts, approaches, and experiences of industrialized countries. This
was done for two reasons; first, industrialized countries possess the capital

and wealth necessary to pursue R&D and implement the most recent technologies to take better advantage of the Internet, and second, as a result of their comparative wealth, industrialized countries have the power to influence and possibly coerce less developed nations to pursue certain strategies (e.g., how will privacy on the Internet be managed?).

How Does This Book Clarify Issues for the Voter or Public Interest Groups?

In part, this book reaches out and addresses such rhetoric as "information highways" and "technology/information haves and have-nots" in the hopes of further crystallizing an emerging social movement around the Internet, computing, networking, and environmental issues. It has become obvious that grassroots lobbying efforts directed at elected officials by their constituents are an important tool of influence in the political decision-making process. Unfortunately, it is not always easy to separate rhetoric from genuine interest in the attempts of new policy to guide Internet practice in a direction desirable to many. The framework, models, and analyses presented in this book can be used as guidelines for most of the Internet-related choices and decisions that we face today. The least one gains from studying this book is some ideas and guidelines about how one may discuss and analyze issues of interest to oneself and others, while trying to participate effectively in the political decision-making process that results in regulations and choices about the Internet.

How Can This Book Balance Such Diverse Objectives as Ethics, Consumer Rights, Profits, and Public Rights?

For some, these ideas may appear to be opposites; in a democratic society, they must be carefully balanced. Pursuing excessive profits or protecting business interests, regardless of the economic and social costs, is not acceptable. We should not use Internet-related things to simply assure reelection (if one is a politician) or to obtain a higher salary through profit sharing (if one is a manager/worker); rather, we should use the Internet to create what is best for the next 50 years. Success or extinction of any social (e.g., church, interest group, etc.) and/or biological organism is accomplished by careful and thoughtful adaptation and change (or genetic mutation). Although it is impossible and not particularly useful to try to predict what might happen if we follow a particular path of Internet choice, we have to think fast and hard about various issues, while trying to reduce mistakes as much as we can. If you submit to this general philosophy, rest assured that in various places you will disagree with the author.

How Can This Book Help in Addressing Institutional Concerns?

It has become obvious that the Internet is becoming a type of institution, but it differs from more traditional forms such as churches in interesting ways. Whereas in church, people may know each other, with the Internet, they may never have met face-to-face, and national traditions, norms, and laws are far less unifying in making this group cohesive than we have seen before. This book provides the reader with examples as they pertain to the emergence of institutional characteristics on the Internet and how different ones may evolve in the near future. Moreover, every chapter links the issues addressed to the institutional theme; hence, institutionalization is a prevalent theme throughout the book and can be found in most places. It is obvious that this latest form of institution without physical boundaries, while evolving in leaps and bounds, represents a challenge for users, researchers, business leaders, and policymakers. Issues of interest and concern to these groups as they pertain to institutional theory and its manifestation on the Internet are discussed here.

How Does This Book Address Related Issues Such as Telecommunication Markets?

Regulation and deregulation of telecommunication markets are of paramount importance to the diffusion of Internet technology. Unfortunately, whereas the United States and Canada have enjoyed deregulation and cheap Internet access through telephone lines and/or cable TV technology for some time, most other users still have to pay communication as well as Internet charges, thereby making accessing the Web from home relatively expansive. Most importantly, commercial use of the Internet by business and private parties in North America has been driven by cheap communication charges due in part to deregulation of telecommunication markets. Unfortunately, innovative and creative use of Internet technology elsewhere might be hampered or at least slowed down by both economic and regulatory constraints imposed on the user by telecommunication, cable, and other markets.

Urs E. Gattiker

List of Abbreviations*

ACM	Association for Computing Machinery (society)
ADAC	Allgemeiner Deutscher Automobilclub (German Automobile Association)
ADSL	Assymmetrical Digital Subscriber Line
AMA	American Medical Association
AOL	America Online
APA	American Psychological Association
ASEAN	Association of South East Asian Nations
ATM	Automatic Teller Machines
BBS	Bulletin Board System(s)
BC	British Columbia
CIS	Computer-Mediated Communication and Information Systems
CISS	Center for Information Systems Security (U.S. Pentagon)
CNID	Caller-Number Identification (also known as Automatic Number Identification = ANI or Caller-ID)
DOD	Department of Defense (U.S.)
EDL	Electronic newsletter/listserver/Distribution List
EICAR	European Institute for Anti-Virus Research

---

*Since these terms are fully explained in the text, this list is provided for easy reference only.

EPIC	Electronic Privacy Information Center (U.S.)
EU	European Union, previously European Community (EC)
GI	Gesellschaft für Informatik (Informatics Society)
GISA	German Information Security Agency
HMO	Health Maintenance Organization
ISDN	Integrated Services Digital Network
IT	Information Technology
LAN	Local-Area Network
NAFTA	North American Free Trade Agreement between Canada, Mexico, and the United States. Following ratification by all three countries, the agreement came into effect on January 1, 1994.
OEM	Original Equipment Manufacturer
R&D	Research and Development
SET	Secure Electronic Transaction
SME	Small and Medium-Sized Enterprises
WWW	World Wide Web

Acknowledgments

My development as a researcher and student of technology and, in particular, the Internet, has been aided by many people. This book provides a welcome occasion for me to express my indebtedness and appreciation. Tina Garstad and Hj Shawyer worked diligently and effectively as editorial assistants during the various stages of this book. They took on a difficult task of forcing me to make my ideas and thoughts clear.

Several of my current and former colleagues at the University of Lethbridge helped to bring the book to completion. I cannot mention all of them, but Shamsul Alam, Lynne Mather, Michelle McCann, Bryan Pratt, Mike Studd, Sajjad Zahir and the former dean, George Lermer, were particularly instrumental in providing guidance for portions of this book on the basis of their own disciplines. A number of their valuable suggestions have been incorporated; they acted as unpaid research assistants and as participants in an informal seminar, as have colleagues from other faculties. All endured many hours of discourse, reacting graciously to my unorthodox questions and unusual assertions. I appreciate all that they have contributed.

A special thanks goes to the staff at the Aarhus School of Business (Denmark), Department of Organisation and Management for providing administrative support, particularly Lis Post. Rita Madsen did all the fine tuning with the indices and with making sure I was not getting sloppy. Many thanks also to Stella Kedoin, *Technology Studies* managing editor for her patience; her extrasensory perception and mind-reading ability were helpful in keeping this project on track. Finally, my new colleagues at Aalborg University were very understanding in supporting my efforts while being patient enough in making me aware of all the cultural and regulatory realities influencing the use of technology not only in Denmark, but in the European Union in general. Many thanks also to Lis Bach who helped in the final stages of this book.

One group to whom I owe many thanks is my students in various sections of the Technology and Change course in Lethbridge, Melbourne, Hamburg, and Aarhus. They patiently reviewed and critiqued my material over the semesters. Some of this help came directly through assessment and expansion of some of the material I gave them; other support came from individuals or groups of students doing class projects, who pursued new avenues for addressing, expanding upon, and solving a potential Internet-related challenge. They were diligent and tolerant—even in situations where it took their teacher a while to understand. I truly appreciate their patience, willingness to share, and hard work.

Special thanks also goes to Helen Kelley, who was an M.Sc. student during 1993–1994. During this process, she became a friend and a colleague, as well as a sounding board. Her enthusiasm, curiosity, and hard work were a motivating factor, which helped my research tremendously. I was also fortunate to have an opportunity to work with Linda Janz, an M.Sc. student during 1995–1997 who spent 1996 in Hamburg with me at the University of the German Federal Armed Forces where we worked together.

Thank you to the reviewers of this book. They helped me find my way during a long journey. Sometimes they told me things I wanted to hear; sometimes, when I needed it, they set me straight.

The first rough draft of this book was written during an extended stay at the University of the German Federal Armed Forces at Hamburg and at the Center for Technology Studies at the University of Lethbridge, Canada. Much of the work on this text was done in my office at home with its beautiful views of Western Canada and at my campus accommodation in Hamburg with a somewhat less beautiful view. The final revisions and rewrites were done in my Danish office at the Aarhus School of Business, and at Aalborg University, in an environment extremely conducive to effective work, both socially and ergonomically speaking.

Finally, I owe a special debt of gratitude to my daughter Melanie, to my friends Inger Marie Giversen, Doris and Dominic Poe, and to Rosemarie, for their unfailing support and their unwavering tolerance of my "downs," especially when the writing process hit a snag. I thank my daughter, Melanie, for making sacrifices throughout work on this project, such as turning down the volume of her radio because "daddy was working," and I thank Inger Marie Giversen for keeping me focused and my friend John Ulhøi, who helped me settle in Denmark in more ways than one.

Urs E. Gattiker
Aalborg, Denmark and Lethbridge, Canada

List of Figures

List of Tables

About the Author

Urs E. Gattiker is Professor at Aalborg University, Denmark and has previously taught in the U.S., Canada, Australia, and Germany. He is a member of Bankinvest's Advisory Board for its BI Technology A/S' IT Venture Fund (http://www.BankInvest.dk) and is also a member of the board of Quiz people A/S (http://www.Quizpeople.com), Vupti A/S (http://www.Vupti.com), B2B Agro Danmark A/S (http://www.B2Bagro.com) among others. Furthermore, he is the head of WebUrb (http://www.WebUrb.net), a group of people dedicated to helping others build their online presence and virtual communities.

Gattiker has written several books including *Technology Management in Organizations*, (published 1990, Sage Publications) and is currently writing on internet and security management.

He was the editor of the book series Studies in Technological Innovation and Human Resources (TIHR) and was the founding editor and currently is editor-in-chief of *TIM-SecurityNews*, an electronic newsletter on information security issues (ISSN 1399–3860, http://Security.WebUrb.dk/ ?SecurityNews).

He has recently edited a book with Laurie Larwood on *Impact Analysis: How Research Can Enter Application and Make a Difference* (1999, Lawrence Erlbaum Associates).

Gattiker's research interests currently include technology change and management, the Internet and World Wide Web, ethics and quality of work-life issues as well as entrepreneurship and high-tech start-ups. In 1994–1995, he served as Program Chair for the U.S. Academy of Management's Technology & Innovation Division and continued as Chair of the

Division from 1996 to 1997 (he chaired the Research Methods Division from 1989 to 1990). He is also a founding member of the Canadian Association for the Management of Technology (CANMOT), now the Innovation Management Association of Canada (IMAC), and the Technology Management Division of the Administrative Sciences Association of Canada (ASAC). He has served in various elected positions for EICAR (Information Security for the New Millennium) (e.g., Program Chair 1999 and 2000), an international association interested in information management and security issues (http://www.eicar.org) and currently serves on the Board of EICAR.

Besides working, Gattiker enjoys reading, hanging out with friends, and discussing various issues besides technology over a good bottle of wine and with some great food being consumed along with it. Moreover, he loves strolling along Denmark's wonderful beaches and bicycling in the countryside through meadows and sometimes wonders what we need all this new technology for....

Setting the Stage
or What It's All About

In this section of the book, I set the stage for the various issues discussed in this book. Chapter 1 outlines how computer-mediated communication (CMC) has been used by people around the world to reach others in faraway places with relative ease. Most dramatic have been the changes experienced with the help of networking computers around the globe resulting in today's Internet. The structure of communication exchanges, skills, and demographics are, however, vastly different across networks (e.g., AOL vs. a local Bulletin Board).

Chapter 2 discusses the rapid growth of the Internet as well as how regulation and deregulation in some markets [e.g., telecommunication in the United States and the European Union (EU)] might have helped or hindered the use of the Internet. In some instances, growth has been facilitated and in others, hindered by regulation and/or the possible lack of regulation in telecommunication, TV-cable, and electricity/power markets. Regulation and/or deregulation is affecting the type of infrastructure offered as well as the services provided. For instance, regulation in the power industry has resulted in the latter using profits from regulated markets to succeed in the deregulated market of telecommunication in many countries. In turn, as chapter 3 outlines, costs incurred by users for products and services provided by suppliers can vary greatly and may in some cases be exorbitant depending on a country's standard of living for the majority of its population. Hence, economic and access opportunities for services and price levels as discussed in chapter 3 are affected by regulatory developments as outlined in chapter 2.

1

An Introduction to the Internet and Computer-Mediated Communication

OVERVIEW

This chapter addresses how the Internet and computer-mediated communication have affected and will continue to affect our lives. Dimensions of media and their attributes are discussed and a model about social complexity and the degree or level of interdependence as it pertains to cyberspace and other media is presented. Additionally, issues of virtual reality are discussed. Finally, how these developments may lead to a new virtual institution is outlined and a model is presented. The chapter is a primer for the uninitiated and is essential for most readers, because everything that follows presumes this knowledge (cf. appendix A).

This book is an attempt to depict some of the development and status the Internet has gone through. Issues, past and future developments, forecasts and ideas are presented and discussed. The book's primary focus is on information technology and end-user computing, global networks and, in particular, computer-mediated communication (CMC). The Internet is not one place or one company. It is a descriptive term for a web of thousands of interconnected broad- and narrow-band telephone, satellite, and wireless networks built on existing and planned communication technology. This infrastructure is a network of networks, reaching out and connecting separate islands of computer, telephone, and cable resources into a seamless web. It connects businesses, governments, institutions, and individuals to a wide range of information-based services, ranging from entertainment (e.g., pay-per-view movies, online music videos), education, and culture to data banks,

3

cyberspace[1] commerce, banking and other services (see also appendix A for further explanation of terms and services available on the Internet).

I do not attempt to determine or predict all social effects of the Internet; I assume that it will have a profound effect on our lives. In most, if not all developed countries, the Internet has led to a merging of communication technologies. Multimedia computer work stations, including CD-Rom and television capabilities for movie, music, and film playing, electronic mail, World Wide Web (WWW) applications, and telephone communication are ubiquitous. Worldwide satellite telephone communication, home automation control, and digital tapes for music and film playing are the trend. Based on today's technology, the Internet is delivered via telephone and power lines, cable connections, and satellite communication. However, if and how these opportunities for delivery of the Internet to each user's home will further develop has been and continues to be, shaped by government legislation and international cooperation. Accordingly, this legislation will have to resolve disputes between telephone, cable and power companies, and media conglomerates. The harmonization of policies between different countries must be achieved and will determine the future of the Internet.

The United States is often referred to as a leader in Internet advocacy, evolution, and utilization. Because of their influence in world policy and the global economy, and the fact that they have one of the largest numbers of computer users and Internet clients per capita, American policy and regulation on the Internet could have universal implications. Moreover, the European Union (EU) is also becoming a powerhouse in world trade and telecommunications. Hence, this book refers to examples and policies from the United States throughout. Comparisons to the EU and other countries is also made.

Although the Internet and advanced telecommunication has made geographical boundaries of lesser importance, the increased cross-border movement of information and data using the Internet has resulted in the United States, the EU, and other countries, as well as associations, making efforts to institutionalize the use, application, and transmission of this communication medium. These norms and roles for organizations and, most importantly, for users are still in flux and rudimentary. Nevertheless, some institutional forms are developing that are outlined as well as discussed and analyzed in this book.

---

[1]The *Webster's College Dictionary* (1991 version) defines cybernetic as the "comparative study of organic control and communication systems, as the brain and its neurons, and mechanical or electronic systems analogous to them, as robots or computers. (… term introduced by Norbert Wiener in 1948)" (p. 338).

In this chapter, I set the stage for this book. Common ground is established, which, in turn, should enhance understanding of the rest of the book without much difficulty. First, how and what type of CMC dealt with is outlined. Second, a short description of the Internet is provided. Third, some definitions about virtual reality and cyberspace are presented. Fourth, institutional matters and concerns are discussed and, finally, some emerging issues are introduced.

CMC AND THE INTERNET

To make the Internet work, CMC and information systems (CIS) are needed that combine the following:

1. Computing that allows processing of content, structuring of participation in, and flows of communication.
2. Telecommunication networks that allow access and connectibility to many others and to varieties of information across space and time.
3. Information or communication resource(s) that range from databases to communities of potential participants.
4. Digitization of content that allows the integration and exchange of multiple modes—such as graphics, video, sound, text—across multiple distribution networks (Rice, 1987).

Table 1.1 suggests the range of CIS but it does not try to be complete, and new ranges for CIS are developed continuously.

Communication media tend to be categorized into simple and mutually exclusive categories, such as mass media/interpersonal, objective/socially constructed, information rich/lean, and natural/technological. Yet, technologies are inherently ambiguous; they can be interpreted in different and possibly conflicting ways, are not always understood, and continue to be adapted and redesigned (Fulk, 1993; Rice, 1992). A multidimensional approach toward conceptualizing media characteristics avoids this binary bias and might help to uncover some of the sublimation of meaning manifested in our perceptions of "familiar" media. Table 1.2 suggests a variety of attributes within four broad dimensions—constraints, bandwidth, interactivity, and network flow—that might be used to characterize media (Rice, 1987; Rice & Steinfield, 1994; Soe & Markus, 1993; see Rice, 1992, for a review of other topologies).

TABLE 1.1

Suggested Range of Computer-Mediated Communication (CMC)
and Information Systems (CIS)

audiotex
automatic teller machines (ATM)

cellular phones & pagers
collaborative systems such as screen sharing and joint document preparation
computer bulletin boards
computer conferencing
conversational and workflow processors

decision support systems with communication components
desktop publishing and document distribution

electronic document interchange (EDI)
electronic mail

facsimile

gophers/World Wide Web
group support systems and other groupware

home shopping and banking
hypertext and hypermedia

intelligent telephone systems
Internet list servers

local area networks

mobile personal communication devices
multimedia computing

online and portable databases
optical media such as CD-ROM and lasercards
optically scanned and networked documents

personal information assistants
personal locator badges
presentation devices such as computer screen projectors

telephone services such as call forwarding or redial until delivery
teletext

video teleconferencing
videotex
virtual reality and cyberspace
voice mail

wide area networks
word processing

Note. This table is adapted from Rice and Gattiker (2000). CIS combine computing, telecommunication networks, information, or communication resource(s) and digitization of content.

6

TABLE 1.2

Possible Dimensions and Attributes of Media
(Face-to-Face to Groupware)

Constraints:

Receiver can identify sender (or anonymous, pseudonyms, aliases)
Have to know receiver's account/name/address/number
Address is fixed to location or terminal rather than person or role
Users have varying read/write/edit/participation modes
Source or centrality of control
Can overcome selectivity
Can maintain privacy
Organizational norms for use
Need temporal proximity
Need geographical proximity
Ease of access to physical location, physical device
Ease of access to and use of interface, commands
Access costs (time, money, energy, knowledge)
Diversity of content available
Diversity of sequencing of content
Can store content (shortterm, longterm)
Limits to message length
Use to transfer documents
Can indicate priority of message
Can store content (shortterm, longterm)
Limits to message length
Use to transfer documents
Can indicate priority of message
Can insure levels of privacy
Can retrieve by indices or browse in random or other order
Can use filtering or allocation processes
Message can initiate other processes directly
Receiver can reprocess, edit for further use
Users can structure flow and privileges
Can easily convert content to other medium

Bandwith:

Analog/digital
Color, images, sound, text, numbers, motion, other senses
Distance

(Continues)

TABLE 1.2 (Continued)

Gestures
Tone, emphasis
Connotation/Denotation
Symbolic aspects or connotations of medium
Social presence/media richness
"Personalize" greeting

Interaction:

Synchronous or asynchronous
(A)symmetry of initiation and response
Speed and type of feedback
Quickness of response by intended receiver
Control receiving pace
Confirm correct receiver, receipt
Mutual discourse
Quick reply feature

Network flow:

Information flow (one-to-one, one-to-few, one-to-many, few-to-few, many-to-many—both of users and of content, such as multiple copies)

Usage domain (human–system, individual, dyadic, group, intraorganizational, community, interorganizational, transnational)
Distortion through overload
Distortion through forwarding edited message
Role effect—can flow be easily controlled
Critical mass necessary

Note. This table was adapted from Rice and Gattiker (2000). The listed attributes are within the four broad dimensions presented earlier and can be used to characterize media.

After looking at this table, one can conclude that CISs have more capabilities than just "overcoming constraints of time and space." The ability to associate or retrieve diverse multimedia content (words, people, events, etc. by terms, concepts, dates, etc.), and therefore the ability to reprocess, combine and analyze information from diverse sources, has profound implications for restructuring society, work, and organizations. For example, while the telegraph allowed people to communicate across time and space, it also allowed railroad companies to collect, associate, and ana-

lyze information from train stations in order to develop effective sched-
ules, routing processes, and billing procedures (Beniger, 1986; Yates &
Benjamin, 1991).

Information is expandable (creates value and new information), generally independent of resources (energy, material, distribution),
substitutable (for capital, labor, materials), transportable (overcomes
time and physical boundaries), diffusive (leaks, not easily appropriable,
requiring other arrangements such as copyright, encryption, monitoring),
contextual (its value depends on the user's needs, perceptions, and abili-
ties) and shareable (i.e., not a zero-sum resource) (Brand, 1987; Cleveland, 1985). Because information is so different from material goods, it is
extremely difficult to measure the benefits of information and CIS using
classical economic criteria (Keen, 1991). An emphasis on efficiency reduces the likelihood of innovative uses of information and CIS because
the time and effort associated with experimentation and learning shows
up only as increased inputs, thus reducing the efficiency ratio (Johnson &
Rice, 1987). In the end, because of the interdependencies among CIS and
other organizational activities (such as training and maintenance), neither
the full costs nor the full benefits of CIS can be readily assessed, if at all
(Keen, 1991).

The digitization of information in and through CIS structures it into a
universal format (bits) enabling it to take form in any communication
mode (text, sounds, video, numbers). It also sublimates CIS from the traditional associations with material structures (i.e., words with books, accurate images with photography, music with records; Brand, 1987;
Mulgan, 1990; Rice, 1987). This detachment removes control of the content from the author, producer, publisher, or custodian (such as a monas-
tery owning the original copy of a book and not allowing copies to be
made). A uniform structuring of media content removes traditional physical structures, thus creating contradictions and problems in the traditional
policies and assumptions associated with physical structures.

What About the Internet?

The media and its possible attributes such as constraints, bandwidth,
interactivity, and network flow indicate that almost as soon as computers
were developed, the need to transfer information between different machines became apparent. Simply using tapes or diskettes to transfer data
from location or machine A to B does work but this requires the physical
transportation from one location to the next. In 1969, the U.S. Department

of Defense began funding the U.S. Advanced Research Projects Agency (ARPANET) that would develop technologies to permit remote research and development sites to exchange information using networks of computers communicating with each other. A major impact of the ARPANET research resulting in today's Internet was the development of the Transaction Control Protocol/Internet Protocol (TCP/IP) network protocol, the language that computers connected to the network use to talk to one another. TCP/IP soon became the standard networking protocol. By the end of 1983, all interconnected research networks were converted to the TCP/IP protocol, and the "Internet," as we now know it was officially born.

The number of hosts has been growing rapidly, whereas the number of domains each associated with a firm, organization (e.g., www.TIM-Research.com), or educational institution has been mushrooming since the mid-1990s. For instance, during the beginning of the Internet in 1983, only 500 hosts were registered; however, by January 1993, there were 1,313,000 hosts whereas by January 1998, the number had mushroomed to 29, 670,000 hosts worldwide (International Domain Survey, 1998).

To a large extent, this growth of registered domains and hosts was in part supported by five developments that all facilitated the diffusion of the Internet and its use by people around the world. These were:

1. The internationalization of the Internet started when ARPANET established connections to England and Norway in 1973, which resulted in the Internet becoming a worldwide phenomenon.

2. TCP/IP became the protocol suite for ARPANET in 1982. The same year EUNet (European Unix Network) began as well.

3. In 1984, JUNET (Japan Unix Network) was established.

4. By 1986, the larger academic community in the United States was given access to the Internet with the founding of the National Science Foundation Network (NSFNET).

5. In 1989, Ohio State University established a relay between CompuServe and the Internet. This meant that the Internet, which had previously been used primarily by researchers whereas commercial networks such as CompuServe were really closed networks now became linked and opened themselves up; research and other users could meet.

6. In 1992, CERN (the European Laboratory for Particle Physics; see also hobbes@hobbes.mitre.org for an extensive chronology of the Internet's history) released the WWW (World Wide Web) graphics-based software that later led to such browsers as Netscape and Microsoft Explorer. Hence, now users could access Web sites with the help of the graphi-

cal browser. The latter in turn provided one the possibility to view papers, products, pictures, and to even listen to soundbites online.

7. By 1995, the diffusion of WWW technology became unstoppable This was reflected by a rapidly increasing number of sites using this technology and, as importantly, more and more users using a graphical browser such as Netscape to "surf" with ease Web sites around the world.

Other important occurrences in the development of the Internet included the end of ARPANET in 1990, when the U.S. Department of Defense thought it had accomplished its ultimate objective by facilitating the establishment of the Internet and thus ARPANET was no longer needed. Moreover, the introduction of Gopher in 1991 by the University of Minnesota facilitated people's searching on the Internet for information and was embraced by many libraries. However, by the mid-1990s, libraries replaced Gopher software with WWW facilities.

These events show that CIS technology enables computers to store, process, rearrange, and transfer information and data quickly and relatively easily. However, only with the TCP/IP network protocol are computers able to exchange information with each other worldwide. Hence, once academic networks were connected to commercial ones (e.g., 1989 Internet node with CompuServe) as well as with others from abroad, nothing could stop the growth of the Internet any longer. Using CIS technology in conjunction with the Internet offers us new opportunities for communication and information delivery, storage, processing of data, as well as entertainment and so on (cf. Tables 1.1. and 1.2).

Cyberspace

The Internet enables its users to transfer information quite rapidly. To do so, various applications are being used such as electronic mail, newsgroups, newsletters, video, and graphics with the help of using a WWW browser and much more. People are able to join chat forums online while also being able to play bridge with the help of the Internet with somebody geographically far away or just down the street. Although we live on our time and space in our "real" lives, the Internet is enabling us to create a parallel cyberspace. Accordingly, some "cybercities" are being created whereby individuals present themselves using the Web and join various groups of people with similar interests. Generally, William Gibson, a science fiction writer, is attributed with coining the term cyberspace. However, in Gibson's work, cyberspace had a negative connotation when tied to the vision of corporate hegemony, urban decay, and neural implants. His descriptions gave a

TABLE 1.3
What is Cyberspace?

Cyberspace is neither a pure pop nor culture phenomenon nor a simple tech-nological artifact. Instead, it is a powerful, collective, mnemonic technology of-fering a computer-generated, interactive, virtual environment of cyberspace. With its virtual environments and simulated worlds, cyberspace is a metaphysi-cal laboratory, thereby providing people with a tool for examining our very sense of reality and the world we live in.

Note. The above definition is linked to cyberspace culture but it may be too narrow (cf. Table 4.2).

name to a new stage in the development of human culture, business, and technology merged into one. The television series, Robocop,[2] also built upon this image of neural implants and life in paranoia and pain.

Cyberspace is a parallel universe created and sustained by the world's computers and communication technologies and is easily accessed via a computer, cable, or telephone modem linked into the system. The com-puter-generated world is faced with some questions about reality. For in-stance, how should users appear to themselves in a virtual world? Should they be seen as one set of objects among others or as a third person that the user can inspect with detachment?

As suggested by the definition in Table 1.3, cyberspace is a mnemonic space offering a computer-generated interactive environment for the user. This environment is kept together with the help of the Internet and is brought to the individual user through cable or satellite communications. In turn, people may wish to remain anonymous or may choose a pseudonym instead of their real identity when taking advantage of the possibilities in cyberspace (Kabay, 1998). In turn, this may result in possible conflicts with people's concepts of ethics, privacy, and freedom that are discussed in chapters 5 and 6.

What is Virtual Reality?

Interactive video games were probably the forerunners of virtual reality be-cause they gave an individual the opportunity to become part of the experi-ence. Nevertheless, there are some important differences. Virtual reality is a process of creating an artificial reality by stimulating the body's sensory

---

[2] A series in which Detroit is full of people in despair and evil characters benefiting from their fear and weaknesses. Robocop, who has the implanted consciousness from a deceased police officer, helps resolve one crisis after another using high technology.

output and makes one's mind and body believe that the image created by the computer is real. Consequently, virtual reality has the potential to involve users in sensory worlds that are indistinguishable from the real world. In addition, virtual reality environments may even merge with the real world. Hence, the body and mind may believe they are experiencing and seeing as well as smelling certain images and tastes, while in reality, one is still standing in one's living room.

Table 1.4 provides a definition of the term, virtual reality. In this book, I am interested in how virtual reality affects and influences the Internet. For instance, technology enables individuals who play bridge on the Internet to find a partner without leaving their homes. Furthermore, virtual reality differs from the interactive video or computer game in at least two ways:

1. When the individual becomes part of the game or situation, his or her consciousness may be altered by adjusting the sensory output, thereby helping one's mind to "register" the virtual environment.

2. The situation may differ every time the individual plays the "game" or virtual situation. Hence, the software may process new information provided by various players and store into its memory banks. This is similar to artificial intelligence where the software develops itself further.

Virtual reality provides the user with additional opportunities to alter his or her experiences or to widen them beyond what is being offered by a video game. Appropriately, virtual reality might help to train fighter pilots to experience the stress and fear of flying over enemy territory or to experience being captured without being physically harmed (Rodgers, 1992).

A virtual theater production may offer actors and the audience new opportunities in experiencing and participating in art creation and expression or performance (cf. quadrant 4, Fig. 1.1). Currently, some of today's TV and radio shows offer viewers the opportunity to participate by phoning or e-mailing in their questions and comments or by being a member of the studio audience (cf. chap. 5). Virtual reality would probably push this a step further but, most importantly, it would build on today's television entertainment. For instance, virtual reality can further increase the degree of sensory experience (e.g., heat, stress) and participation for the viewers. To illustrate, some theaters include the audience in the play; virtual reality offers far greater participation of the audience and interaction with the main actors. However, going to the Hamburger Staatsoper (Hamburg State Opera) is a social event and although one may want to see the opera, sometimes a person may also want to be seen by others! Also, social interaction before and after the play and during intermission is an important part of going out to the opera. Hence, virtual reality may be able to provide the person with many things but not all, as this example about the opera suggests.

VIRTUALITY AND COMMUNITIES

The previous section outlined such terms as virtuality and cyberspace (cf. Table 1.4). In the context here we are particularly interested in virtual communities and the Internet. Virtual community as a term is defined in Table 1.4 and is a term that has been mentioned extensively in the business world and by marketing experts. Needless to say that a virtual community is based on virtual reality; this is to say that a virtual community may never be exactly the same as a social community. In the latter case, people may share culture, experiences, language, and much more, including interests such as making a community or neighborhood a better place to live and raise chil-

TABLE 1.4
Virtual Reality and Virtual Community?

Virtual reality is a process of creating an artificial reality by stimulating the body's sensory output, thereby making one's mind and body believe that the virtual reality created by the computer image is reality. The ultimate objective is to give the user a feeling of being present in an artificial environment.

Virtual reality allows the user to experience the sensation of being present in an environment. One perceives objects and images as being present. Presence is defined as a person's sense of being in an environment. Presence is grounded in telepresence, which is a person's experience of presence in a simulated environment with the help of a communication medium (i.e., computer technology). Computer technology provides the person with the feeling, look, and sounds of another environment.

Virtual reality is based on several devices, each addressing a particular sense. For instance, a head-mounted display shows the user visual surroundings while the computer, with the help of software adjusts the images as the user moves and travels through virtual space.

A virtual community is based on virtual reality, that is, it is usually an artificial social construct whereby people with similar interests (e.g., growing a rose garden) meet and exchange ideas, messages, and communicate (e.g., pictures, sound, video files).

Note. The above is linked to cyberspace culture, as defined in Table 4.2. Virtual reality may also provide the individual with the opportunity to simply become a recluse and find a refuge from day-to-day life in an artificial environment (cf. Fig. 1.1).

dren. Hence, a social community may in part be virtual (e.g., e-mail, fax, and telephone) but there has to be a "real" or "social" side to it to make it a social community (see also Table 8.2). Hence, social communities are grounded in social networks and thus are based on social interactions between individuals sharing culture, living space, and possibly values (see also chap. 4).

Evolving Virtual Communities and the Institutionalization of the Internet

One could assume that virtual communities may have certain structures that, in turn, facilitate interaction and working together (see also Fig. 8.1). At this stage, we could for simplicity's sake compare virtual communities to buildings that have structures in the form of beams, interior walls, toilet facilities, and so on. Similarly to comparing social communities with virtual communities, comparing older cities with newer ones indicates that the latter were built with the car in mind (e.g., Los Angeles), whereas older cities were built assuming that most people would walk, use bicycles or public transportation. Accordingly, the highways and roads are built differently and determine if the city is made for pedestrians or cars instead (number of streets, their wideness, parking facilities, and so on). Whereas architects and designers may agree on building a house or a street after having carefully solicited and integrated into their designs input from various stakeholders, organizations, and institutions, social and virtual communities evolve.

Whereas Los Angeles was a city designed and built with the car in mind, many much older European cities were built when, aside from a horse buggy, people used to walk instead. Accordingly, sometimes a city's road infrastructure is hopelessly overused and thus results in traffic jams as is the case in London. Or, cities may take drastic measures to reduce individual traffic by car to cope (e.g., Bologna, Singapore, and Copenhagen) by restricting access, charging fees, or developing many more pedestrian zones, one way streets, and limiting parking making driving into the center unattractive. The question is how do social communities evolve compared to virtual ones. Are they also constructed and if so, does a comparison similar to the one made earlier between older versus newer cities make sense?

Traditionally, we understand social communities as evolving and being in part based on people sharing culture, language, experiences, and folklore, and experiencing and interpreting events such as natural disasters similarly (for a detailed list of characteristics, see Table 8.2). In the virtual community, meeting one face-to-face is not possible although membership can change by having people sign on or off as they wish. In a building organized as a co-op, members may know each other by having worked for the association (e.g., as volunteer) or by having met and talked during the co-op's last barbecue. But even in a housing co-op, community spirit may be limited to owning part of the building and changes as well as the developments affecting one's quality of living and equity value of one's share in the building. Strength of social ties also affect a social community. Weak ties (acquaintances) may be more useful if they are nearby (close geographical proximity), whereas strong ties (family members) may remain useful even if they are far away (e.g., across the country). Thus, the proximity of community resources may affect the closeness of a community.

Institutionalization.   People in organizations or institutions may not necessarily agree on what the organization or institution should be like. With buildings, once they are built, they remain the same until a major remodeling may be done a few years down the road (e.g., adding nice washrooms to hotel rooms in a building from medieval times). Accordingly, houses and office buildings today look different than the ones built in the 1960s or 1930s. With organizations, their structures may change due to fads or fashion. Also, organizations evolve and change over time depending on the environment (see contingency theory) and the people working in the organizations.

The Internet has certain similarities to an organization insofar as it consists of people. An organization's design is "... along various lines of people among social positions that influence the role relations among these people" (Blau, 1974, p. 12). In the context of the Internet, various groups of people and organizations with various positions and needs as well as roles affect the relationship between these various parties. For instance, the Internet has removed geographical boundaries, thus cultural differences and similarities may become an issue for users of the Internet (Gattiker & Willoughby, 1993). Other differences occur between academic or research users and business as well as consumers all having different needs. In some instances, universities may have started to use the Internet in some ways that are similar to business (e.g., trying to reach a larger audience for their products and services). In other instances, universities have undertaken commercial or semicommercial activities themselves (e.g., science parks, joint ventures that use patents held by the university and/or one of its researchers).

The previous discussion would suggest that because of its many diverse user groups, the Internet is not necessarily a very cohesive institution or organization but instead, represents users with diverse interests, values, cultures, and languages. Accordingly, this multicultural organization is difficult to manage as governments have discovered when trying to regulate it (see also chaps. 4 and 8). But these issues also apply to virtual communities as this book shows. Although there may be a community that has been established for a while and that can claim to have a stable membership, it is unlikely to be cohesive for starters simply because geographical proximity does not exist. However, a virtual community can develop and change in its focus, interest, or issues being addressed and shared by members, similar to a building. Moreover, as the following section illustrates, interdependence and structure of a community may play an important role in how it may function.

DEVELOPING A CONTINGENCY MODEL

So far we have discussed how computer-mediated communication provides people with new opportunities as far as processing, transferring, and distributing information as well as creating content. Although technology may impose some restrictions as outlined in Table 1.2, opportunities are vast and the use of the Internet by private users, organizations, governments and firms for diverse purposes is growing rapidly worldwide. However, the rapid increase in the use of CIS technologies and the Internet in particular is also influenced by social complexity and interdependence. Doing an activity on the Internet such as typing an e-mail message is socially less complex than playing chess with a partner in cyberspace. Moreover, composing e-mail offline makes one less dependent than playing chess online, because in the latter case, losing the connection might force one to restart the game.

In addition to complexity and interdependence, diffusion of the Internet technology and the structure imposed on its use could also affect what we can and cannot do on the Internet. Before we can speak of the Internet as an institution with well-established rules and norms, however, it is necessary to discuss in a schematic way how this development may have been influenced by or contingent on various forces, which are outlined below.

Social Complexity and Degree/Level of Interdependence

Playing a video or computer game has a different degree or level of interdependence than being a part of a sports team or theater group. Each group differs based on its social complexity and interaction. Being a part of a team, either in sport or theater, requires an individual to attend training and rehearsal sessions. An individual is dependent on others to show up for rehearsal or practice. Failure to attend may result in social pressure to change one's behavior accordingly. Playing a video or computer game or reading a book is distinct from the previous scenario because one can do either whenever one feels like it, assuming one has access. Hence, no scheduling is necessary; it is an individual activity.

Based on social complexity and the degree or level of interdependence, we should be able to develop a theory of fit by outlining a technology or mode contingency theory. Here, mode means a logically coherent pattern of structure and process matched to a level of CIS/social complexity determinism. The structural elements of these modes are defined in terms of:

1. specialization, acquiring of different skills by individuals to perform a number of activities both at work and in private life;
2. transferability of these attained skills to other activities;
3. interests, existence of clearly identified and defined opportunities of entertainment that provide enjoyment opportunities within one's life; and
4. lateral mobility, entertainment transferability representing the individual's own choices to do different activities and acquire new skills required, in part, to use new technology during one's spare time.

Process is defined as the coordination mechanisms used by the individual to facilitate the fit between him or herself and the entertainment. Figure 1.1 illustrates the relationship of choices made by an individual regarding entertainment opportunities and technology-related changes. The degree or level of interdependence and social complexity are shown to have an influence on the individual and others involved in that person's life.

Social complexity is lowest when the person engages in a simple activity, and highest when the final product/outcome is evolving as part of a process over time involving several individuals (e.g., a team playing a game). Although Fig. 1.1 presents the degree of interdependence and social complexity in four Quadrants, each of the two dimensions is not a dichotomy but a continuum instead.

Quadrant 1 demonstrates that the social complexity and the degree or level of interdependence (social and technical) is very limited. A person may borrow a book (e.g., from the library), but thereafter, they can read it

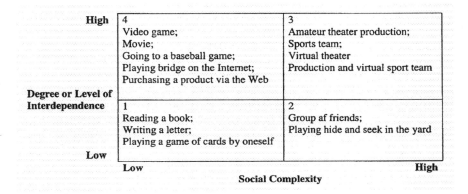

High	4 Video game; Movie; Going to a baseball game; Playing bridge on the Internet; Purchasing a product via the Web	3 Amateur theater production; Sports team; Virtual theater Production and virtual sport team
Degree or Level of **Interdependence**	1 Reading a book; Writing a letter; Playing a game of cards by oneself	2 Group af friends; Playing hide and seek in the yard
Low	Low	High

Social Complexity

FIG. 1.1. The level of interdependence and social complexity on the Internet. Both dimensions represent a continuum and not a dichotomy. Reading a book is generally a single person's activity, whereas an amateur theater production requires various functions to be performed (e.g., producer, costume, and set design) to secure that the final output (opening night) will be a success. Accordingly, social complexity and interdependence between functions and people is highest in quadrant 3.

Often, less social complexity is implied if one performs an activity by oneself, whereas being part of a process such as rehearsing for a play requires several people to be involved and thus it becomes more complex.

anywhere without being dependent on any other resource (i.e., except for light to be able to read by—unless the book is in braille). One is also not required to interact with other individuals (social complexity).

At the opposite extreme in Fig. 1.1, activities existing under the more munificent and benign conditions of Quadrant 3 are marked by high social complexity and degree or level of interdependence. This individual experiences a complex environment and is dependent on others (e.g., fellow team members). Social interaction during an activity is complex and sometimes infrequent, considering practiced automatisms (e.g., playing a certain strategy to score a touchdown in football).

In Quadrant 2, social complexity is great when children or friends are playing a game in the backyard, whereas interdependence is limited. Accordingly, a friend may join or leave the game anytime without interrupting the game or play. He or she can be replaced with another peer joining the group. However, social interaction and role playing (the bad vs. the good people) are a more complex form of entertainment. Conflicts about role interpretation or participants' behaviors (e.g., being responsible for a foul or a turnover in football) necessitate the individual using social skills.

Quadrant 4 is a relatively stable situation for the individual, because no social complexity seems apparent. When an individual plays a computer or video game, he or she is dependent on the software and electricity making the machine, and thus game, perform the tasks they are supposed to. Of course, this assumes one does not play the computer game via the Internet with several partners around the world. Even then, however, social interaction is limited.

Although a book reader may not be a hermit, he or she could be a recluse trying to limit social interaction as much as possible or could be less outgoing than the person who does a lot of activities while being a member of a group. Certain CIS-related activities could reduce social complexity; nonetheless, other skills may increase, such as working interactively as well as having technology skills required to use CIS technology effectively (Gattiker, 1990a). Figure 1.1 does not suggest that CIS reduces social complexity automatically or that a dichotomy is apparent. Social complexity is higher when one reads a book and then discusses it with friends (e.g., bible study). When several individuals play a video game or surf the Net together, one of them must "drive" the event by using the keyboard or mouse. This requires social interaction. Nevertheless, it appears that technology can help in reducing social complexity, as past developments would suggest (cf. quadrant 4). Even playing bridge using the Internet does not provide the individuals with as much information (e.g., nonverbal communication) and social complexity as playing and chatting with each other in person. Accordingly, technology as represented by a computer game or by using the Internet to play bridge (quadrant 4) is increasing one's dependence on the technology to function properly while reducing social complexity. However, it also offers the individual the choice to do certain things while staying at home, thereby possibly increasing convenience.

Recent developments in the virtual reality domain offer entertainment with sound, picture, and the experience of social interaction in a virtual space (see quadrant 3). We don't know if the social complexity of a virtual reality theatre production will be as high as a live theater production. For starters, each actor is in a different geographical location. The opening and closing night parties might occur in the virtual coffee shop. Moreover, a virtual theater may permit people to remain more anonymous than if they were to rehearse in the same space for several weeks or months. To illustrate, an individual may be able to appear different than he or she is in reality. Using a pseudonym or anonymity options may permit individuals to represent themselves quite differently than their immediate family or friends perceive them and experience them on a daily basis in real life. Also, understudies could be asked to take on the role for which they prepared more often in virtual theater than in live theater.

Diffusion and Structure of Internet Communities/Spaces

CMC and the Internet not only have to be studied as far as interdependence and social complexity is concerned but, as importantly, before we are able to understand the institution better, we also need to schematize the relationship between its diffusion and structure. With the term diffusion I mean the number of people or social groups being part of a subculture or a particular network (e.g., AOL). Structure represents the rules and regulations as well as the norms adhered to by users. The greater the rules and regulations, the more unified a group of users becomes. Hence, using sophisticated filters to prevent certain messages from being passed on (e.g., the ones containing pornographic content), or having chat rooms monitored by system administrators to make sure that participants adhere to etiquette all represent rules and regulations and control trying to impose institutional characteristics on participants. Accordingly, failing to adhere to these rules may result in a user being excluded from the institution, that is, no longer being given access to the services and products offered by a provider.

Governments have also tried to further unify the Internet and make it into an institution adhering to certain rules, regulations, and behaviors. In some instances, governments have tried in vain to filter information and possibly keep it away from their citizens if it is deemed politically incorrect from their vantagepoint (e.g., People's Republic of China). But most of these efforts have been futile, similar to trying to impose people to follow one architectural design school to build their private houses would not be successful either. Moreover, because geographical boundaries no longer have a meaning, imposing unified schemes of norms and rules has become a regulator's nightmare as later discussions show (e.g., chaps. 4 and 5).

Figure 1.2 further illustrates these issues using the level of diffusion (low vs. high) and the level of structure (low vs. high). Again both these dimensions are not dichotomies but represent a continuum instead. Figure 1.2 assumes that the roles and relationships on a local bulletin board of Star Trekkies (*Star Trek*: famous TV show with several motion pictures) from

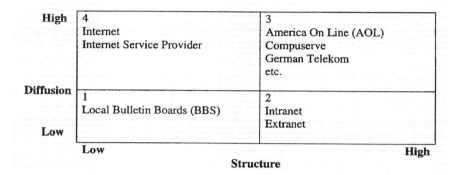

FIG. 1.2.   Diffusion and structure of the Internet or information technology networks. Both dimensions represent a continuum and not a dichotomy. Also, it is obvious that a local BBS reaches fewer people than AOL. Hence, diffusion or reach of Internet netiquette practiced by the BBS members or information distributed through a BBS is lower than is the case with AOL. Additionally, it is likely that a BBS structure (e.g., rules and regulations) imposed on its users is lower than what AOL requires of its users in order to retain user privileges.

the local high school might have few rules and might be accessed by a few high school students only. In contrast, America on Line (AOL) reaches millions of people but has much structure about content and users' behavior on the Net when using AOL services or accessing the Internet through AOL (see also latter parts of the book). Similarly, an organization's Intranet has very strict safety, security, and privacy rules as well as rules about user behavior and so on. In contrast, accessing the Internet through a service provider reduces the rules to a minimum and unless somebody complains, consequences are minimal, especially, because norms and cultural constraints of country A may not apply to user in country B (e.g., United States vs. Lithuania).

Quadrant 4 also suggests that a person may use a direct link, with one's own server and line, to the Internet in a country where the laws are very general about information technology and distribution. Similarly, phone sex companies often provide their services through countries where the laws are lax (e.g., Dutch Guyana) until they are tightened due to outside pressure, and then the firms move on. When the client calls a number abroad, the call is connected through again to the country the person is calling from, where a human provides the service. This allows the firm to charge higher fees and circumvent local laws that might restrict the use of phone numbers that cost the person a lot of money.

Demographics, Culture, Code of Conduct, and Netiquette. Table 1.5 outlines the characteristics of the quadrants as represented by the various user groups. Looking at user demographics, the differences between these systems are relatively small for gender, age, and education. Nevertheless, CIS users tend to have higher levels of education than the average population. For language, it is obvious that as soon as the reach increases, English becomes dominant. Culturally, systems have different values and norms but the greater the reach, the more likely U.S. culture influences the characteristics of the system's culture. Computer skills indicate that Internet and BBS users are probably the most skilled users as far as computer know-how is concerned.

As Table 1.5 also indicates, differences across the four quadrants can also be found as far as a code of conduct is concerned. Whereas BBS might have only a very general one, AOL and Compuserve represent systems (quadrant 3) with relatively strict codes of conduct that are again influenced by the English language and culture (original is in U.S. English), even though each subsidiary may have a translated and somewhat localized version for its clients. Enforcement of such codes is likely to be strict on a company's Intranet system as well as on a commercial service provider such as Compuserve and even Internet access providers (e.g., EuNet, AT&T, Tele Danmark) have become careful to avoid legal problems.

If looking across the quadrants and Netiquette (see Table 1.5), the differences are such that the smaller the reach, the more tight the net. Within a country, people use e-mail and the telephone in certain ways, reflecting culture and language (institutionalized norms require one to provide certain information to the called party when he or she picks up the phone; if one fails to do so, one may not be connected or may simply be perceived as impolite). Again, using the Internet, the Netiquette is not very clear whereas national standards may be obvious to netters; communicating between users in various countries may make finding the appropriate Netiquette more difficult. The emoticons reflect similar trends as Netiquette, that is, although there are some symbols used around the world, most users do not know them all and may therefore rarely use them besides a happy face :-) or

Table 1.5

Four Types of Structure: Approaches to Formalization and User Characteristics on the Internet

	Quadrant 1 *BBS* *Low Reach and Structure*	Quadrant 2 *Organizational Intranet* *Low Reach and Low Structure*	Quadrant 3 *AOL/Compuserve* *High Reach and Structure*	Quadrant 4 *Internet* *High Reach and Low Structure*
User demographics				
Gender	depends on topic	majority male	majority male	majority male
Age	depends on topic	usually skewed toward younger users	usually skewed toward younger users	usually skewed toward younger users
Education	average or above	average	above average	above average
Language	local	local/sometimes English in international firms	local language but provider supported/sold content is primarily in English	local language but internationally mostly English
Type of culture	uniform, tightly knit	dependent on organizational rules, norms, beliefs, and myths	differs across user-groups from different countries (e.g., Germany vs. U.S.) but overall heavily influenced by U.S. culture	relatively international but heavily influenced by Anglo-Saxon mannerisms

(Continues)

TABLE 1.5 (Continues)

	Quadrant 1	Quadrant 2	Quadrant 3	Quadrant 4
	BBS	Organizational Intranet	AOL/Compuserve	Internet
	Low Reach and Structure	Low Reach and Low Structure	High Reach and Structure	High Reach and Low Structure
Computer skills of end-user groups[a]	above average	average or slightly below average	average or slightly above average	above average
Code of conduct				
Yes/no	unlikely, except for some generally understood rules	large organizations increasingly have one; medium and small-size firms are less likely to have one	each local subsidiary has one that user must abide by; AOL U.S. code is, however, used as template for local codes	no general code except for some unwritten rules whose violation may result in such consequences as flaming; Internet provider may have a code
Enforced	possible but likely very loosely enforced	enforced if caught; consequences may range from minor to being fired or sued	enforced strictly; consequences are losing system privileges	not strictly enforced, violators are politely or less politely (e.g., messages returned with profanities included) reminded of their wrongdoing; local authorities may try to enforce some laws …

(Continues)

TABLE 1.5 (Continued)

	Quadrant 1 BBS *Low Reach and Structure*	Quadrant 2 Organizational Intranet *Low Reach and Low Structure*	Quadrant 3 AOL/Compuserve *High Reach and Structure*	Quadrant 4 Internet *High Reach and Low Structure*
Netiquette	is clear to all members because tightly knit group	is usually obvious to all users within the firm	may differ between services/countries (e.g., love chat room vs. medical advice forum); nevertheless, general norms apply	no clear rules or norms; however, nationally accepted standards of politeness are usually adhered to by all users; North American mannerism dominates
Emoticons	whatever is understood by fellow BBS users	rarely used in business context	generally understood and accepted emoticons are used; however, sexual and other undesirable content is not tolerated lightly by discussion group moderators or system administrators	generally understood and accepted emoticons are sometimes used; emoticons may differ between countries

Note. This table outlines how the degree of reach on the Internet and level of structure (see also Fig. 1.2) may affect the formalization and user characteristics in the various CIS systems.

[a]Computer skills of most end-users including those who may not have Internet access. In Fig. 1.2, quadrant 3's user skill levels are explained by the user friendliness of these systems, thereby enabling the end-user with limited skills and understanding of technical issues to easily access such a system ("the whole family").

a hug ( ). However, certain systems like AOL may enforce moral norms that prevent any sexual undertones or words to appear in discussion groups or on Web pages. For business users, such symbols are rarely applied because most communication is supposedly dry and task/assignment focused, whereas on the BBS, new icons may be used whose meaning is known only to that group of users.

Institutionalization and Advertising.   Figure 1.2 and Table 1.5 both illustrate institutional characteristics albeit of a type of institution loosely connected in the case of the Internet. Nevertheless, one of the reasons for the rapid growth of the Internet has been its limited structure (rules and regulations) that has permitted BBS, Internet service providers as well as intra and extranets to start up shop all over the world with a few mammothtype organizations, such as AOL, in between. However, although the latter are large, they have been struggling either to become or remain profitable for quite a while, and whether they will dominate the business in the future is unclear and probably unlikely. Competition is increasing, especially because content, as offered by AOL, is supplied more and more by other services for free, that is, by having advertisers pay for Internet users getting a service for free (e.g., http:// www.excite.com or http://www.bigfoot.com).

This indicates that the Internet is going in a similar direction as once occurred for newspapers and television, that is, advertisers are picking up a large part of the tab in order to make these products available to consumers at a low price or for free. Accordingly, further institutionalization of the Internet will mean that proprietary networks such as AOL must offer their clients unique and special content to keep them. This will likely become harder and harder because advertisers are discovering many new ways to reach their current and potential customers via the Internet (see also chap. 6).

Institutionalization, Structure, Social Complexity, and Interdependence of the Internet.   Figures 1.1. and 1.2 as well as Table 1.5 provide the reader with some structure about important dimensions of the Internet affecting its level of institutionalization as well as the social complexity and interdependence of users of the system. Subsequent chapters in PART I of the book focus on issues affecting interdependence for users based on regulatory matters as well as costs incurred to use the Internet. Part II outlines how cultural diversity and different interpretations of ethical and moral issues as well as perceptions of e-commerce may all further hamper institutionalization efforts undertaken by various groups but especially by governments such as the U.S. government.

SUMMARY AND CONCLUSION

This chapter provides a short introduction into the topic being discussed in this book. Specifically, terminology used throughout the book is introduced and defined for the reader's benefit, and some challenges and opportunities for the Internet are outlined. How CIS technology works and the characteristics needed to describe media is presented in Fig. 1.1. There is a short discussion about cyberspace and virtual reality. The potential impacts of the Internet on the political process, through lobbying by interest groups using CIS effectively to advance their agendas, are also outlined.

Important in this context is also that until 1989, when CompuServe was linked to the academic and research network, Internet growth between the research community and the world outside (e.g., business) was limited. That is to say, CompuServe users were unable to communicate with people working in academic institutions and government agencies because a gateway between the two systems was lacking. However, after such a link was established, Internet's growth for private and business use as well as for educational application was unstoppable around the globe.

As discussed, using the Internet and CIS (cf. Fig. 1.1) for certain activities can increase or decrease social complexity for the individual. The choosing of one activity over another may, in part, be due to an individual's preference. As well, technologies may increase the degree or level of interdependence. Without resources and certain pieces of technology, an individual cannot participate in an activity on the Internet. He or she requires a telecommunication connection to a host computer on the Internet, a computer with the appropriate software needed to play the game, and the electricity to run the hardware and software. Finally, an ever more expensive commodity, namely time, is needed to play the game! In 1997, AOL reported that its users spend nearly 15% less time watching television than the U.S. average (about 7 hours less per week); instead, they spend time online (e.g., surfing the Web, participating in chat rooms, doing e-mail, etc.; Internet users spending, 1997).

Figure 1.2 describes how reach and the level of structure can result in differences based on social demographics, code of conduct, and Netiquette/emoticons. To simplify, we can say that the wider the reach of a system, the weaker its possibility for enforcing its structure (i.e., rules and norms). Having users adhere to unified norms is difficult unless the user is part of a worldwide service (AOL and Compuserve) whereby violators of codes of conduct or norms/rules may just lose system privileges. Further parts of this book illustrate how the difficulty in establishing structure and institutional norms can in part be explained by regulatory, cultural, ethical, and other differences exhibited by users from around the world.

The chapter indicates that the Internet provides us with a vast array of opportunities. It challenges how we communicate, entertain ourselves, and perform our work. The rest of this book investigates these challenges and opportunities in depth. Implications for research and decision makers is also outlined.

2 Regulatory Developments and Internet Policies

OVERVIEW

The Internet has been developing rapidly in the last few years and its use worldwide continues to grow. Accordingly, regulatory developments in telecommunication as well as legal concerns require attention to assure a smooth use across borders and continents. Here, some of these regulatory developments are outlined and examples illustrating these points are provided (cf. Appendices 3 and 4). As the discussions illustrate, institutionalization of the Internet is hampered by the lack of harmonization of regulatory efforts across countries.

Policies and regulation are an important vehicle for any institution or organization to establish rules and norms that people can live by. Moreover, such norms and rules facilitate the interaction and exchange of communication or data between various parties. Berger and Luckmann (1967) argued that social reality is a human construction, being created in social interaction. The process by which actions are repeated and given similar meaning by self and others is defined as institutionalization.

Regulation and deregulation of markets result in new rules, norms, and behaviors by users to communicate with each other. In this chapter, my focus is on how regulation and deregulation in telecommunication may affect the Internet and, most importantly, how laws about encryption and safeguarding information may limit social actors or may raise their capacities to protect vital information. Here, laws may both create new realities and typifications and affect social reality for parties using the Internet. However, several divergent regulatory developments in the United States and the EU indicate that various actions and behaviors are interpreted differently by politicians and regulators, thereby resulting in different outcomes for Internet users, which could hamper globalization of electronic commerce (e-commerce) and the Internet's growth into an institution.

REGULATION AND DEREGULATION

Government regulation affects the way one does business. If one regulates the consumer marketplace according to Ramsy (1985, p. 353) three issues are raised:

1. Why do consumers need protection?
2. When ought governments intervene?
3. How ought governments to intervene?

In most instances, governments try regulating a market to eliminate some undesirable behavior, to protect some segment of society from perceived harms, or to provide information necessary for making informed decisions. Most users require telecommunication or cable technology to connect to the Internet from home or the office. Hence, regulation about telecommunication and cable TV does affect Internet users. In deregulated markets, anybody can offer Internet, telephone, or cable TV access through the phone or the cable TV connection to the house or even by using another technology (e.g., electricity lines). Naturally, deregulation is supposed to improve the price and quality or service relationship thereby providing the client with better value for his or her money.

Industrialized Countries and Developing Countries

In an industrialized country, deregulation of telecommunication markets may have been implemented for some time (e.g., England started with deregulation in 1982). In a developing country, however, the main issue for the government and its people may still be to provide drinking water and electricity to most neighborhoods. Hence, deregulation of either telephone or cable services may not be an item on the government's agenda or in the public's mind until more important infrastructure issues have been taken care of. Accordingly, if we discuss Internet regulation and policy, we need to first address some basic issues dealing with access.

Access to Telecommunication Technology and Cable.   Access to a telephone is a simple matter for a person in an industrialized country. For instance, over 90% of Canadian households have telephone service. However, of approximately 800 million people living in the industrialized world, 400 million enjoy telephone service from home. In contrast, of the 5 billion people living in the developing world, only about 200 million (4%) enjoy such a service (Bauer, 1995). For example, about 70% of Africa's population live in villages without either electricity or telephone connections. Moreover, half of the world's population has yet to make a telephone call, and in many African countries, only about 1% of the people enjoy ownership of this century-old communication technology. This indicates that whereas in some countries we talk about the Internet and multimedia world, in others, the dirt path is still the only way to get around!

A similar story can be told about access to cable TV services. Again, approximately 80% of all Canadian households already have access to cable (Angus & McKay, 1994) and similarly percentages can be found for most industrialized countries. Naturally, unless electricity is available in a dwelling, telecommunication technology usually can not be used except for a radio often powered by a manually driven generator.

This illustrates that in an industrialized country, deregulation of telecommunication and cable as well as of electricity markets (e.g., United States) is of importance to regulators, businesses, consumers, and politicians, whereas in many developing countries, providing electricity and hopefully subsequent access to phone service to villagers may override regulatory concerns. In the latter case, service may eventually be available, but as the following section shows, costs for the technology might be prohibitive for many groups of society.

Costs and Standard of Living.   Providing access to the Internet might be a concern to a family in an industrialized country whereas people in the developing world are more concerned about food and shelter. For instance, it was estimated that in 1995, an African family of 4 in the Gauteng area (Johannesburg/Pretoria region of South Africa) had a gross income of approximately R 1,600, which is substantially lower in the rest of South Africa. This means that Internet access costs would take up about 10% of these families' disposable income (after taxes and housing costs; see Table 3.1). Because it is unlikely that the majority of the country's population will have a computer at home, as well as access to telecommunication services, the challenge is to provide inexpensive access to the Internet for the individual without severely taxing the public system.

This illustrates that in many developing countries, current economic conditions as well as the level of infrastructure in most neighborhoods makes it unlikely that the majority of citizens will gain Internet access in the near future from home. Here, a viable alternative might be to create access possibilities with the help of public resources and institutions such as libraries for the majority of the population (see also Access and Cost section).

Market Deregulation as an Opportunity for the Internet.   What the previous illustrates is that both access to technology options and the costs in relationship to living standards are important variables affecting the diffusion of the Internet, or its use by private citizens and organizations alike. But even in an industrialized country where deregulation of various markets has begun, in some cases since the early 1980s, certain developments are hampering benefits that were supposed to accrue from such efforts. For instance, in Britain where deregulation started in the early 1980s and has been brought in slowly over the last 14 years, a 3-minute call to the United States has fallen by 70% between 1996/1997 and 1998 alone. Unfortunately, prices for local calls, where competition has yet to affect British Telecom, have remained stable.

The first and biggest threat against deregulation paying off in better service and lower costs for clients or fairer competition, unfortunately, is the forming of near monopolies or oligopolies in the cable, electricity, and telecommunication markets. In most instances, cable or local telephone service providers as well as electric utilities offer their goods or services as single supplier to each household within their "territory." This means that most households can choose their electricity or cable services from one supplier only. Accordingly, further restricting business such as having cable and local phone services supplied by one firm cannot be to the advantage of consumers. Unfortunately, regulators often seem to miss the early signals. For instance, the Swiss Competition Office did not step in when, in 1996, Swiss Telekom (the past monopolist for telecommunication services) was permitted to take over Rediffusion, the largest cable service provider with over 60% of the market in Switzerland. Similarly, when the Swedish telecommunication firm Telia took over the Danish cable provider Stofa, it became the dominant force in that industry for Denmark. In addition, the firm is also a major telecommunication player in Denmark offering long distance and cellular services. German Telekom has always been the largest provider of cable services in that country. If we can learn from the car industry, it seems obvious that innovation, better quality, and hopefully lower prices are most likely to occur in a market segment (e.g., small cars) where there are at least three firms offering competing products in a geographical market. In the Swiss and German example, unfortunately, the dominant telecommunication provider is also the leader in the cable business. One does not need to know much about economics to understand that this surely will not facilitate competition for communication services (e.g., TV, Internet, and telephone) and thus price decreases for customers will be smaller than otherwise possible, as other markets have shown (e.g., Finland).

A second threat comes from firms that were or still are able to cross-subsidize from regulated markets into deregulated ones. For instance, in Germany RWE, the largest electric utility, in combination with another six energy firms, can cover 70% of the country with their own fiber optic network. Unfortunately, this network was paid for by profits generated by overcharging their clients for electricity, thanks to their monopoly in that market in certain regions of Germany.

A third threat stems from a scenario whereby a telecommunication provider is using profits from regulated markets and income received from private customers paying fixed line and toll charges to compete, providing services cheaply to organizational users. Once a fiber optic cable has been laid down and paid for by private customers, additional Internet traffic (e.g., by leasing a standard line) can be carried without much, if any, variable costs being incurred by the telecommunication firm. This scenario has happened all over Europe whereby former government-owned telecommunication firms are often not sure how to calculate prices fairly for permanent line connections to the Internet. In turn, they may be offering permanent line connections to organizations below full costs simply to remain competitive or to gain market share against new private start-ups. New telecommunication start-ups have been shown to require three times fewer people to generate the same amount of profit than their large competitors previously owned by their respective governments. Accordingly, fair pricing based on fair cost and investment accounting seems necessary to assure a fairer competitive environment for all market participants.

The fourth threat against fair competition occurs whereby a public organization or corporation received subsidies to build an infrastructure, of which parts are now being used to launch new ventures in deregulated markets. For instance, in Germany and Switzerland, the national railway has developed its own infrastructure for telecommunication, which before deregulation was used internally only, thereby helping the railway to avoid the high costs charged by both country's telecommunication provider. These two systems built with public funds permitted both railways to join private telecommunication consortiums, offering highly competitive services to private and corporate customers. Here, government subsidies and taxpayers' money for developing and maintaining infrastructure for trains is being used to embark and offer services in newly deregulated telecommunication markets. If the railway and the electric utility join forces, suddenly a deregulated telecommunication market is to a large extent dominated by firms who have amassed cash and infrastructure in regulated markets.

Using Shadow Pricing.   The competition threats two through four as previously outlined are basically dealing with problems stemming from the leftover of regulated markets. Hence, a firm having previously operated in a regulated market or still doing business in a regulated market (such as electric utilities and railways) while also competing in deregulated markets (e.g., telecommunication) may have an unfair advantage. Although a public organization entering new markets or trying to expand market share should be commended, careful deregulation seems necessary to avoid undesirable cross-subsidies. Economists would suggest the use of shadow pricing.

Shadow pricing is a concept used by economists to determine the price for a product and/or service for which the firm may not have a clear idea or formula for calculating realistic costs. Here, economists suggest we look at the market price for the product or service we need. In the case of the railways and subsidized infrastructure for its own telecommunication system, costs paid today to develop such a system may be used as a basis to determine costs charged to a new venture and, in turn, may provide revenue for deficit-generating activities such as passenger train service.

Deregulation May Work.   A truly deregulated telecommunication market should provide most users the choice to choose their Internet access and its actual use from several providers offering various types of services for different prices. Providers may include but not be limited to telecommunication, cable TV, and electric utilities as well as satellite communication (e.g., Iridium). All these firms have the necessary infrastructure to provide Internet services at competitive prices to private households in industrialized countries. However, cross-subsidies from regulated into deregulated markets must be avoided. If we have learned anything from telecommunication market deregulation, it is that more than one provider should offer services in local and national or international markets. For instance, in England, deregulation of telecommunication markets in 1982 was gradual and prices for long distance calls did not come down until the early 1990s. Because local calls were still regulated, price reductions were not forthcoming until the late 1990s (*A Map of the Future*, April 4, 1998). Naturally, this also goes for cable-TV service and electricity where most households do not enjoy the option of choosing between several suppliers. If a consumer can choose from several suppliers, quality of service improves while prices tend to come down. Within the next decade, the rapid bundling of services and the possible synergies between power, cable, and telecommunication as well as satellite communication services should help consumers and organizations to get better value for their communication dollar.

For developing countries, however, it will take some time until consumers and organizations will benefit from a better price and service or quality relationship for communication (i.e., voice and information using cable and/or telecommunication technology), even in large cities. One reason being that in such cash-strapped nations, the huge investments required to build the necessary infrastructure is hard to come by. Moreover, if governments succeed in getting the necessary funds, the limited use of such infrastructure (i.e., because a large portion of the population simply can not afford it) does not permit the economies of scale necessary to encourage private investors to offer competing services. Even satellite communication networks may not provide these countries with an economic alternative except for a few of its wealthier citizens.

ACCESS AND COSTS

The regulatory framework determines to a large extent how easily parties get access to the Internet at what price. For instance, if the telecommunication market is deregulated and prices are relatively low compared to the consumer price index, as is the case in North America, access to the Internet is relatively easy for private and organizational users. However, if many households have neither electricity nor phones, and/or the standard of living enjoyed by the majority of the population is low compared to the United States, things may be very different. Here, government may undertake policies to make Internet access for citizens a possibility by using facilities at public institutions. Some of these options are now discussed.

Public Library Access

North American public libraries have provided citizens with easy and inexpensive access to literature, music, and movies. Accordingly, individuals are able to borrow books, cassettes, movies on videos, and much more to be enjoyed at their respective homes with their families. A North American or a Scandinavian library is probably quite different from a South African or Argentinean library. For starters, North American and Scandinavian libraries have existed for some time developing into institutions from which communities expect certain services. Moreover, in industrialized countries, libraries' infrastructure is usually quite good if not excellent. For instance, a city such as Aalborg, with less than 150,000 people, has a library offering literature in several languages including newspapers and magazines, music, movies, and naturally, Internet access. Unfortunately, like most public libraries in Europe, Aalborg's library is not open evenings or on Sundays.

Internet access enables library users to check their web-based e-mail for free because neither Internet access through the library nor web-based e-mail costs the user (e.g., web-based Hotmail in the United States is free for users worldwide). If communication charges are not incurred by the user due to the library having a leased line, even slow web-based e-mail requiring one to be online all the time while composing and reading e-mail is of little consequence. The only drawback with such an approach is that the individual has to visit the library to make use of the Internet and, therefore, has access only when the library is open.

Another approach taken is where libraries actively engage in the creation and management of Freenets and local Bulletin Boards (BBS). However, some have raised criticism about such activity (see section on Freenets) especially because taxpayer money might be used to subsidize services that are offered at competitive prices in North America. Even in a developing country, only a small and usually much better off group of citizens would be able to use a library-supported Freenet from home. Once again, this would question the justification for using public funds to support such a project benefitting a very small group enjoying the luxury of having a personal computer at home.

Nevertheless, we can conclude from these discussions that public funds can be used effectively by libraries to establish the *virtual library*, whereby patrons have the opportunity to search the Internet for information they wish to obtain. Additionally, clients can also check their Web-based e-mail messages on servers that offer such services for free to private users whereas the library simply provides the terminal and Internet access through its leased lines. Such an approach is probably the most economical way of using public funds to help people without access from home to take advantage of the Internet and, most importantly, become literate and acquire skills that are becoming a prerequisite for many jobs (e.g., Gattiker, 1995) in developing and industrialized countries.

This does, however, also demand that the library develop the concept of the virtual library further. In turn, patrons will have the opportunity to access information and catalogues at the library during open hours. At most universities today (e.g., Aalborg University, Denmark or University of Lethbridge, Canada), more and more of the library's holdings are available online and the university's Intranet permits users to even read articles from scholarly journals online or print a copy if so desired. Users who are not on campus are still able to log on and search the library's many catalogues and databases. Accordingly, the virtual library permits users to access rare or highly specialized collections around the clock 365 days a year. Public libraries have also begun to become more virtual but in many countries besides North America, their efforts are still dismal in comparison to university libraries (e.g., Germany, Italy, Spain).

Educational Institutions and the Internet

Another viable policy option for governments trying to assure that their fellow citizens become Internet literate is to provide the necessary funds for schools and universities to enable these institutions to provide access to the Internet at a nominal fee or better for free to students. Recent developments, however, suggest that some governments are trying to get their universities to charge their pupils for access to the Internet (e.g., state of Idaho). In some instances, educational institutions in the province of Alberta (Canada) may provide 10 megabytes of e-mail space for free whereas modem access may cost anywhere from $7.00 to $15.00 a month. Although such charges are not exorbitant, many students will not subscribe to such services in order to save the money for other purposes such as going out on Friday nights. Here, it looks as if neither students nor governments and universities are behaving smartly. Other options such as restricting daily access via telephone and cable modems (e.g., 2 hours), or during popular times (e.g., 1 hour from 5 p.m.–midnight, Mon–Fri) would help reduce the demands on universities' infrastructure.

Another development that seems to tax university's computer systems and modem/cable access from outside campus is the use of Web-based e-mail programs. Although these might be cheap or even free for the university library to offer to students, such programs require students to be on-line to compose and read e-mail. Other software, such as Eudora Light (also free) would allow students to compose and read e-mail offline. Moreover, mail from the university's computer could be transferred to the student's personal computer at home while subsequently being automatically deleted from the mainframe system, in turn, reducing storage space on the institution's computers drastically.

In most industrialized countries, between 20–60% of households do have a computer and the numbers are rising rapidly. Many, if not most, of today's students arrive on campus at their first semester with a computer offering them Internet access. Accordingly, cheap modem or cable access for students offered by their universities will encourage them to use their own computer hardware and software that, most importantly, reduces the demand for computer work stations requiring supervision on campus.

This illustrates a North American phenomenon, whereby more and more institutions of higher learning are trying to recoup some of the costs incurred due to students and staff accessing the Internet from home. Hence, it appears that shortsighted cost accounting and successful lobbying by private Internet access providers may have resulted in some North American universities charging students for home access, which could be considered unwise from an educational strategy point-of-view. In contrast, in countries where telecommunication charges are metered (see Table 3.1), Internet access from home for students, staff, and faculty is not an issue. For instance, students in Denmark use the Internet primarily on campus to avoid telecommunication charges. Similarly, students in developing countries might not have their own computer or telephone access from home, again making the providing of modem access to students an unlikely cost issue for universities.

The issues described here indicate that a dichotomy exists between countries. Hence, in some countries, unlimited Internet access from home for free for students is being challenged due simply to universities being forced to pay part of their own way with fees. In other places, providing Internet access on campus for all students is still being implemented while Internet use in teaching remains scarce. This also indicates that the *virtual campus* is possible, where students have access to the library and other sources including the World Wide Web (Web) with course materials being offered through the Web, and lectures, discussion groups, and group assignments also being conducted online. However, while in North America, virtual campus may mean the student participating in learning from one's home, in Europe, due to high telecommunication/cable access costs to the Internet (see also Table 3.1), it means learning from a satellite campus usually located in a rural area. In the latter case, students in more remote locations may attend such a satellite campus, which provides them with the necessary computers and Internet access to participate in the virtual university.

Cooperation Between Institutions.   Cooperation between learning institutions should also be encouraged. An example of how learning institutions can work together is The Western Cape Schools' Network, founded in South Africa in 1993, and is the first and largest independent school networking organization in Africa. This network is proof that schools, with the help of nongovernment organizations, businesses, and volunteers, can assure accessibility for teachers and pupils to the Internet at low cost. Currently, the cost of this network starts as low as $100.00. This does not include telephone charges, which can range from approximately $5.00 up to $80.00 for a dedicated line (locally only), and feed into the network's hub at the University of Cape Town. The ultimate goal of this network is to one day have these institutions lease lines only to the next school, to avoid trunk call charges altogether.

Universities and colleges are ideal gateways for schools throughout a region to gain access to the Internet. To make such a strategy successful, however, the regulated telecommunication markets may have to be encouraged to permit schools to lease dedicated lines to these institutions at reasonable prices.

The Internet can also be used to foster collaboration and learning between students from schools and universities in different countries. For instance, INTOPIA (International Operations Simulations), a computer-based business simulation software, was used with the help of the Internet between student groups in Lund (Sweden) and Münster (Germany). The game enables students to execute various business transactions, to communicate with each other in English using e-mail. Moreover, competing groups, with the help of video technology, can meet each other toward the end of the game. The use of business simulations in training is on the rise with the Internet further facilitating the collaboration between universities in order to improve the educational experience of students.

Freenets or Savenets

The previous discussion illustrates that various public institutions can be used to provide people Internet access. Although getting Internet access from a library is not as convenient as from home, nevertheless, it illustrates that without any additional costs, people can gain access to the Internet from such institutions at very low cost to the public purse. The school and university examples also demonstrate that pupils learn to master this new technology if collaboration between educational institutions is encouraged, thus avoiding turf wars. Finally, if short-sighted cost accounting or profit mentality is replaced with strategic vision, universities can be encouraged to provide inexpensive Internet access to their students from home. In turn, this will definitely be an important factor for further encouraging the use of the Internet in educational delivery to students by their staff (e.g., virtual campus).

Nevertheless, in some cases Internet access may be a problem for private households due to standard of living, infrastructure (e.g., no electricity), or not having access to a computer. In a developing country, the library may also not offer Internet access either. However, in most industrialized countries, if certain economies of scale have been attained, markets become attractive for new firms to enter and offer products and services to meet a growing demand. One response to solving the access problem while achieving attractive economies of scale in Canada, the United States, and New Zealand has been for some public groups to form local "Freenets," whereby users benefit by getting subsidized access to the Internet. Freenets are restricted to noncommercial use (Cunningham, 1997). Freenets increase the use of the Internet by private users and may actually create the economies of scale to encourage private firms to start offering Internet access at reasonable prices.

With Freenets, however, taxpayers pay for these services directly in the form of a grant or indirectly by supporting a library that in turn may provide space for the hardware required to run the Freenet. Some Freenets use local universities as gateways, thereby being subsidized by educational funds. Unfortunately, it is usually in the larger population centers where Freenets have been established, instead of the smaller centers where they are needed. The question is if government should subsidize these Freenets in markets where competitive environments already exist; that is, where suppliers have an incentive to reduce rates in order to gain market share (see Table 3.2).

Table 2.1 presents an alternative to Freenets in the form of Savenets whereby government subsidies are limited to a particular time frame. Similar to a Freenet, a Savenet is a not-for-profit organization supported by volunteers' time as well as by equipment donations made by individuals and organizations. To avoid another creation of an "entitlement" mentality by voters, governments are encouraged to limit dollar support per user for each

TABLE 2.1

Savenet

	Basic Fees	WWW—Fees per Hour	Telecom Charges Low Tariff During Evening/Weekends for 5–6 Minute Call	Average Charges for 10 Hours online and About 60 Messages in and/or Outgoing per Month	Average Monthly Charges in USA $ Using the UBS Purchasing Parity Ratios1 (Approximate Numbers Only)
1) Electronic mail	U.S. $6.00[a]				
2) FTP/Telnet	U.S. $4.50				
3) WWW	U.S. $4.50	U.S. $15.00 (Services 1–3)	U.S. $0.18/local	U.S. $29.00	U.S. $29.00[b]
4) Video & Sound	U.S. $4.50	U.S. $19.50 (Services 1–4)[c]		U.S. $33.30	U.S. $33.30
5) Telephone communication	U.S. $6.00	U.S. $25.50 (Services 1–5)[d]		U.S. $39.30	U.S. $39.30
6) Cable TV & services	U.S. $17.00	U.S. $56.30 (Services 1–6)		U.S. $56.30	U.S. $56.30

[a] A small community may decide to simply offer a Bulletin Board that offers local information and access to other members of the board. Intermittent Internet access is provided, that is, instead of full 24-hr. connectivity; hence, electronic mail is uploaded and downloaded from an Internet access point 2–3 times a day. In 1995, Mauritius offered such a service for its university staff via a South African gateway in Johannesburg.

[b] This figure includes Cdn. $14.00 for local telephone charges. Accordingly, for a Canadian or U.S. customer, these charges would drop to the Internet access fee only, approximately Cdn. $14.00 per month with taxes. Because Canadian and U.S. consumers can get this type of Internet access at this or even lower price levels from private suppliers if they live in cities with more than 50,000 people (metropolitan and surrounding areas), today's SaveNet option is not necessary and, in fact, may represent a misallocation of scarce public funds.

[c] We estimate that by the year 2000, more and more retailing of music, videos, and other material will be done via the Internet. Hence, the CD with one's favorite music may be downloaded directly from the music company's WWW site (e.g., Sony and Time Warner). This will require additional bandwidth, for which the user will be asked to pay for as soon as this becomes a more prevalent way to sell music and videos. However, the consumer will likely save on the purchasing price, because the distributor saves costs as well (e.g., retail store's margin). Hence, one purchase a month may more than offset the additional fee to be paid.

[d] Using the telephone option will again require additional bandwidth that the Internet access supplier must provide to give consumers this option. Additional costs will have to be paid by the consumer. The price advantage will still be substantial even with this surcharge, because the consumer may pay local access charges to the Internet provider only, instead of the long distance charges one incurs with the telephone company by using the telephone for making calls to one's friends around the world.

year over a 5-year period. The Savenet I envision requires users to pay based on the type of service chosen. Ultimately, the Savenet objective is simply to jump start the use of the Internet in smaller communities, thereby achieving the economies of scale that, in turn, encourage private suppliers to provide various services at highly competitive prices.

The Savenet suggestion offers the user unlimited access for the fee proposed for each category using 1998 and 1999 price levels. Today, categories 1–3 have been implemented on a large scale in industrialized countries. The pricing structure presented is similar to today's costs for telephone services, that is, besides basic services telephone companies offer various add-on services (e.g., caller display/identification, voice mailbox) and additional costs are incurred by the consumer based on his or her choice. Because cable companies are entering the local and long distance phone market in the United States and England, there is no reason why cable TV and related products and services cannot be provided through the Internet assuming that bandwidth is made available (see Table 2.1). A government should choose to support a Savenet per user/annum, but this fee should be minimal because its thrust is simply to create enough users to encourage competition by several private Internet access providers. Subsidizing this process for private consumers does not appear to be a wise public policy decision. In fact, it creates another dependency where none is needed.

For a developing country, a Savenet is not really an option because the majority of households need electricity and telephone access first, before the Internet becomes a concern. Moreover, it seems more opportune to spend money on public institutions such as libraries, which, in turn, can provide users with the possibility for finding information on the Internet. The way I envision viable Savenets makes them similar to cooperatives, whereby users organize themselves to offer alternative services in a cost-effective manner but without government subsidies.

Savenets or Freenets have to be discussed because wherever private users' access is yet limited to the Internet, various parties suggest the use of Freenets. Simplistically speaking, nobody would think of offering free telephone service to certain groups of citizens. Hence, it makes little sense to offer free or heavily subsidized Internet use to people who might already be better off than others in a society. Such need for Internet access can be more economically served by using other public institutions' facilities such as libraries without creating another dependency on public coffers where none is needed.

LAW AND INTERNET: LOOKING AT EUROPE

The previous sections illustrate that certain policies about deregulation will affect who will provide what infrastructure to private and organizational Internet users at what price (see also chap. 3). Moreover, the collaboration of various institutions and the availability of ways for users to take advantage of the Internet are other options to help both people with limited resources and those lacking the necessary infrastructure (e.g., electricity) to gain inexpensive access even in developing countries.

Once large groups of citizens have the opportunity to gain access to the Internet, however, governments try to regulate a market to eliminate some undesirable behavior, protecting some segment of society from perceived harms, or providing information necessary for making informed decisions. Most regulation and laws today were developed within the context of the traditional media of print, radio, television, and telecommunication technology. The Internet and WWW differ from these established media and technology in that they remove traditional barriers of time and distance, thereby enabling the marketer to develop databases of product information and enabling users to selectively obtain information from around the world. In this section, I first discuss how countries' efforts to protect their citizens' privacy while making communication and electronic commerce viable and secure has resulted in some institutional efforts across national boundaries. The EU and other European countries (e.g., Switzerland) are used as examples in how difficult it is to develop legislation and laws making electronic commerce a viable and safe option for all participants across national borders. A harmonization between NAFTA and the EU, or OECD guidelines and local legislation seems advantageous but as the following discussions show, we are still far away from such developments.

Regulation of the Internet

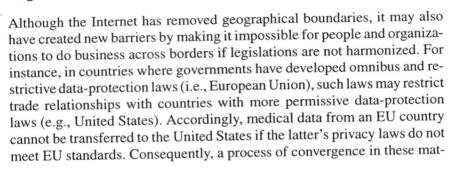

Although the Internet has removed geographical boundaries, it may also have created new barriers by making it impossible for people and organizations to do business across borders if legislations are not harmonized. For instance, in countries where governments have developed omnibus and restrictive data-protection laws (i.e., European Union), such laws may restrict trade relationships with countries with more permissive data-protection laws (e.g., United States). Accordingly, medical data from an EU country cannot be transferred to the United States if the latter's privacy laws do not meet EU standards. Consequently, a process of convergence in these matters is evolving (cf. Bennett, 1992) whereby EU legislative efforts lead to emulation by other government body legislators to ensure the easy exchange of data between organizations within and outside of the EU. The EU's own efforts are resulting in the harmonization of privacy legislation across member states. For instance, failing to harmonize environmental laws to meet EU standards, but instead just having directives, led to Germany being taken to court by the European Commission in January 1997. Failure to comply and harmonize its environmental laws was costing Germany millions of marks for every passing day, because the court approved the case in favor of the plaintiff.

Nonmembers of the EU are suffering from new legislation (such as the environment) in the form of *penetration*. Penetration pertains to the event whereby certain adjoining EU states, such as Switzerland, paid the costs for changing their privacy legislation and its administration to meet EU guidelines by mid-1998. Moreover, telecommunication markets in Switzerland were being opened for competition using a similar timetable as the one followed by the EU. *Elite networking*—the convergence of knowledgeable individuals from different countries to discuss common interests and concerns—will help to spread the EU's ideas concerning Internet legislation. This may result in compatible legislation for all nations involved and help to assuage the conflict between the EU and nations like Switzerland (cf. Bennett, 1992, p. 5).

Regulation at Work.   Switzerland has recently made an effort to achieve legislation against damages caused by altering programs and/or data by third parties (i.e., releasing a virus[1]). This is a good example to illustrate the process of institutionalization of cyberspace. To illustrate the need for adding such an article to the criminal or penal code, Swiss legislators discussed Italian and Swedish developments on this matter. Hence, a process of convergence was used by outlining, to a parliamentary commission, the need of such an article to catch up with neighboring states' efforts (Frigerio, 1995). Instead of simply emulating Italian and Swedish efforts, the Swiss made an attempt to improve legislation. For this, elite networking was applied by informing Newsgroups and newsletters on the Internet discussing these matters about the legislative discussions and efforts undertaken in Switzerland. In turn, this knowledgeable and interested audience

---

[1]For a definition of this term and an explanation of various types of computer viruses, the reader is referred to Appendix A.

(e.g., including Interpol specialists, hackers and virus experts) was invited to provide feedback in the form of comments, suggestions and criticism, which was collected to refine and improve the proposed legislation (Frigerio, 1995).

It is interesting to note that Article 144bis (Damaging of Data) focuses on damages instead of outlining the penalty for a particular offense (e.g., writing a computer virus, see Italy or England against hacking). The new law assumes that a virus is written to manipulate data that nearly always causes damage to the other party; hence, it can be prosecuted under this new law. Similarly, hacking or cracking a computer is a violation of this law. Accordingly, this new article is inclusive by focusing on damages attributable to a previously performed activity. Hence, the Swiss article 144a tries to avoid the notion of cyberlaw and as Karnow (1994) suggested, "legal rules peculiar to the electronic communications context—makes no more sense than printlaw, newspaperlaw, movie law, T.V.-law, shopping center law, videogame law or, indeed, washing-machine law" (p. 7).

The Swiss example shows that some degree of convergence evolved whereby legislative efforts in Italy and other countries led the Swiss authorities to emulate these. Elite networking was used to improve legislation by getting the advice from knowledgeable individuals from around the world. However, the challenge will be to harmonize Swiss law with the EU and NAFTA. Similar to the EU's privacy legislation, the Swiss ultimately need to assure a harmonization with EU states as far as their damaging of data legislation is concerned, thereby assuring a further penetration of their efforts beyond their local "shores."

Penetration of EU privacy legislation is illustrated by Bill 68 in Québec (Canada), which was approved in 1993. It represents Québec's anticipatory response to EU efforts, and its hope that by tightening its own regulation, it may exceed EU standards. Most important, data exchange and, thus, trade between the EU member states and Québec firms is not disturbed. No place in North America, with the exception of the province of Québec, has a law that adequately protects personal information held by the private sector. The United States and Canada have federal Privacy Acts but, unfortunately, they only protect personal information in the hands of the federal government. Some Canadian provinces do have privacy protection at the provincial and municipal level but, besides Québec, none offer particular protections outside the realm of government efforts. This shows that a common set of laws further signifying institutionalization of cyberspace is a challenge for legislators. Furthermore, as the following section shows, whereas geographical boundaries may not exist for users, they do exist for legislators.

Encryption and Digital Signature

As much as the previous discussion indicates that countries have searched for various solutions when it comes to regulating the transfer and exchange of data and information between parties in two countries, similarly, encryption and digital signature issues have been dealt with in various ways. If harmonization is not accomplished between EU countries, problems can occur. For instance, the North American Free Trade Agreement (NAFTA) protects a company from Mexico investing in the United States and vice versa. However, the treaty's 11th chapter gives foreign investors the right to demand international arbitration, claiming expropriation of their assets, if they think that any government action unfairly harms their investment or crimps their ability to trade. In turn, legislation applied to local companies may be deemed illegal when applied to firms based in another NAFTA country. One of Ethyl's (U.S. firm) fuel additives was banned in 1997 in Canada on environmental grounds and has been in its home country for some time. Ethyl is claiming that this amounts to expropriation and a NAFTA panel of experts is hearing the case. Here it could be that one technology for encrypting data may not be used legally in the United States but outlawing it in Canada may be challenged using chapter 11 of the NAFTA treaty. Without harmonization, this problem can hardly be resolved satisfactorily.

Digital Signature

An important application for public key cryptography is the "digital signature," which can be used to verify the integrity of data or the authenticity of the sender of data. In this case, the private key is used to "sign" a message, whereas the corresponding public key is used to verify a "signed" message. Public key cryptography offers the benefits of confidential transmissions and digital signature in an open network environment in which parties do not know one another in advance. This development allows for broader applications of the cryptographic mechanism, and this, together with increases in computer power and decreases in computer price, has moved cryptography into the private sector domain.

Public key cryptography and digital signatures play an important role in developing global information infrastructures. Much of the interest in information and communications networks and technologies centers on their potential to accommodate electronic commerce; however, open networks such as the Internet present significant challenges for making enforceable electronic contracts and secure payments. Several different methods exist to sign documents electronically, varying from very simple methods (e.g., inserting a scanned image of a handwritten signature in a word processing document) to very advanced methods (e.g., using cryptography). According to the European Commission (October, 1997), electronic signatures, based on "public key cryptography," are called digital signatures and are widely considered as crucial for a variety of applications like:

- official communication with public institutions (e.g., calls for tender, exchange of application forms, identity documents, tax declarations, transmission of legal documents),
- contractual relations in open networks (e.g., electronic commerce and financial transactions),
- for identifying or authorizing purposes (i.e., in order to be certain of the identity of correspondents or their attributes, such as having the authorization to log into a computer system or accessing a restricted part of a network),
- in closed systems (Intranet), and
- for personal purposes.

Naturally, these applications are rather extensive, and misuse or abuse of the technology is possible. Accordingly, application issues must be addressed and firms as well as governments are still grappling with these issues. Without eliminating potential fears of users concerning secure and safe use of such signatures, their use will be limited as the following section outlines.

Application of Digital Signatures.   There is a tremendous potential for fraud in the electronic world. Transactions take place remotely, without the benefit of physical clues that permit identification, thus making impersonation easy. The ability to make perfect copies and undetectable alterations of digitized data complicates the matter. Traditionally, handwritten signatures serve to determine the authenticity of an original document. In the electronic world, the concept of an "original" document is problematic, but a digital signature can verify data integrity, and provide authentication and nonrepudiation functions to certify the sender of the data. If a document itself has been altered in any way after it has been "signed," the digital signature will so demonstrate. Similarly, once a document is signed with a cryptographic key, the digital signature provides proof that the document was signed by the purported author. In turn, the sender cannot easily deny having sent the document or claim that the information has been altered during transmission (cf. OECD, 1997).

Legal Questions With Digital Signatures.   The implementation of digital signatures as a cryptographic mechanism to support authentication and nonrepudiation security services seems to offer a technical solution to a legal problem. However, it must be mentioned that there are some legal questions that need to be addressed and, unfortunately, they are addressed differently in various countries. The following aspects are currently being discussed regarding legal concepts behind digital signatures, and requirements on form and procedures:

1. Does a digital signature meet legal requirements?
2. Is a digitally signed document recognized as evidence in court?
3. Does a "Declaration of Intent" have a legal value?
4. Are there technical solutions to make sure that users sign a document in the version that is actually visible on their screen?
5. Does a digital signature prove that a particular person actually signed a given document?

The last point especially is of some concern because conventionally, a person signs a signature by hand. In the electronic world, however, the technology would permit a third person—authorized or unauthorized—to sign a document, if this person is in possession of the private key ("undisclosed delegation").

Cryptography can also provide technical solutions for the protection of intellectual property in digital form. For example, a digital signature together with a verifiable timestamp can give authors some control over their work by tying an electronic document to the issuer and ensuring that the document is not modified without detection. The same technology can be applied to ensuring the authenticity and integrity of documents archived electronically (OECD, 1997).

The Need for Action Within Europe.   The earlier sections on cryptography and, especially, digital signatures suggest that countries have to move beyond the basic agreements made under the OECD's umbrella (OECD, 1997). Some member states of the European Union have already proceeded to develop detailed regulations for digital signatures. Germany has released a law on digital signatures (Bundesministerium fuer Bildung, Wissenschaft, Forschung und Technologie, October 8, 1997). France has adopted a new Telecommunications Act (Loi N° 901170, 29 December 1997) that is being criticized by various groups, including the French Parliament's Commission for Post & Telecom Public Service (CSSPPT) (Thorel, 18 December 1997). Italy adopted a law on the use of electronic documents and contracts (Council of Ministers, August 5, 1997). The English government has launched a Public Consultation on the regulation of Trusted Third Party encryption [(TTP) uk_crypt]. The Dutch Government has created an interdepartmental task force (Staatscourant nr. 54, 18.3.97). Denmark and Belgium are also preparing draft legislation on digital signatures [http://www.agoraproject.org/]. The Swedish government organized a public hearing in June 1997.

Although the development of a clear framework is welcomed, the very divergent legal and technical approaches that have already appeared constitute a challenge. Moreover, the absence of any legal environment in some EU member states—although possibly justified—might constitute a serious barrier for communicating and doing business throughout the EU using digital signatures. This will undermine the free circulation of digital signature related products and services within the EU's domestic market, while hampering the development and expansion of electronic commerce. If ecommerce should be stimulated, then obstacles to digital signatures and free circulation of information must be of highest priority. As well, facilitating the use of digital signatures across national borders requires a common framework at the EU level. Such a framework is urgently needed and should be put in place at the latest by the year 2000.

Confidentiality and integrity of data are facilitated if encryption technology is applied, thus increasing privacy of data being exchanged legitimately between two parties. In the next section, I discuss how encryption policies and legislation may facilitate or hinder organizations' efforts in protecting privacy of their client data.

Export Control Measures.   Concerns over foreign threats to national security have been the primary motive for export controls. Whereas countries want to protect their own military and diplomatic communication through encryption, the objective of export control is precisely to deny similar benefits of cryptography to foreign opponents. Therefore, export controls are in general designed to prevent international proliferation of certain encryption technologies.

Under the Wassenaar arrangement on export controls for conventional arms and dual-use goods and technologies replacing the Coordinating Committee for Multilateral Export Controls (COCOM), a group of 28 countries applies export controls to encryption products.

Within the European Union, the Dual-Use Regulation (December, 1994) establishes a common framework for exports of dual-use goods [Council Regulation (EC) 3381/94, December 19, 1994]. This regulation sets up a community regime for the control of exports of dualuse goods (i.e., civil and defense applications) by establishing a list of goods covered by the regulation. Accordingly, certain encryption products may only be exported on the basis of an authorization. In order to establish an internal market for dual-use goods, such export authorizations are valid throughout the EU.

Moreover, according to Article 19 of this Dual-Use Regulation, member states exercise a license procedure for a transitional period and for intracommunity trade of certain particularly sensitive products. For the time being, this also includes encryption products. This means the regulation obliges member states to impose not only export controls (i.e., controls on goods leaving Community territory) on dual-use goods, but also intracommunity controls on cryptography products shipped from one member state to another.

This Dual-Use Regulation, however, does not fully specify the scope, content, and implementation practices of national controls. Consequently, a large variety of domestic licensing schemes and practices exists. Divergence between domestic and EU regulations as well as others can lead to distortion of competition.

Domestic Control Measures.   Law enforcement authorities and national security agencies are concerned that widespread use of encrypted communication will diminish their capacity to fight against crime or prevent criminal and terrorist activities. For this reason, in several EU member states, consideration is being given to how their encryption policy could develop in the future. This has led to national and international discussions about the need, technical possibilities, effectiveness, proportionality, and privacy implications of such a regulation.

Existing Regulation Within the EU and the OECD.   Whereas export control measures are internationally widely applied, up to now, domestic control of encryption is quite exceptional. In fact, currently only one member state of the European Union, France, applies a comprehensive cryptographic regulation. The Prime Minister (in practice the SCSSI, which stands for "Service Central de la Sécurité des Systèmes d'Information") submits provision, export, and use of cryptography to a simple declaration that the cryptography can have no other object than authenticating parties or ensuring the integrity of information, otherwise it must have prior authorization. This law is currently being modified (Loi N° 901170, 29 December 1997). Although there have been discussions in other member states, only England has so far launched a public consultation on the regulation of TTPs for the provision of encryption services [but not for use of encryption (DTI, 1997)].

The international picture is quite similar. Looking at the OECD countries, besides export controls, there are basically no domestic regulations implemented. In the United States, where up to now no domestic regulation is in place, there is an intensive debate on several legislative initiatives. In taking up the developing debate on this topic in some OECD member countries and trying to avoid obstacles to international trade and commerce resulting from divergent national policies, the OECD has adopted guidelines for a cryptography policy.

Regulation About Use of Encryption.   Regulation of use would mean to rule the use of encryption without an authorization as illegal. Alternatively or additionally, supply and import of encryption products and services could be brought under an authorization scheme. Authorizations would either be denied or granted under certain conditions, for instance, to use only weak encryption or to sell only approved software. These conditions are scaleable to satisfy any perceived needs of law enforcement and national security agencies.

Such regulations could limit the use of encryption. In addition, divergence between regulatory schemes might result in obstacles to the functioning of the internal market, in particular for free circulation. If an encryption software company, which can freely develop its products in its home country, must comply with specific technical or legal requirements in other member states, this company has to produce at least two, if not more, different versions of its encryption software. The same situation occurs if enterprises want to offer crossborder encryption services.

Today, nobody can be totally prevented from encrypting data, especially criminals or terrorists that can use encryption for their activities, for three reasons:

1. Access to encryption software is relatively easy, for instance by simply downloading it from the Internet.

2. It is difficult to prove beyond reasonable doubt that an accused has sent an encrypted message without prior authorization. Electronic communication on open networks is not like an end-to-end telephone conversation, where people can be identified by, for instance, their voice.
3. Encryption is also possible using steganographic methods. These methods allow one to hide a message in other data (e.g. images) in such a way that even the existence of a secret message and thus the use of encryption cannot necessarily be detected.

As a result, restricting the use of encryption could well prevent law-abiding companies and citizens from protecting themselves against criminal attacks. It would not, however, totally prevent criminals from using encryption. Most of the few criminal encryption cases that are known and used for justifying governmental regulation of encryption concern "professional" criminals or terrorists. However, it is unlikely that such parties would be stopped from using this technology even if regulations would prohibit them from doing so (cf. Denning & Baugh, 1999).

Some agencies have suggested the use of TTPE (see Appendix A for a definition of the term) to reduce misuse of encryption. The purpose of TTPE systems is to preserve the ability of the intelligence and law enforcement agencies to access and decrypt information if necessary to fight crime and terrorism. With the help of TTPE, procedures for disclosure to such agencies of encryption keys are established. However, such an approach faces two major challenges:

1. the sheer volume of keys to be stored over time assuming a wide use of encryption technology by consumers, medical and other organizations as well as governments;
2. transnational concerns whereby a person or organization's communication is decrypted in another country by a law enforcement agency that would not have been given these keys under a more democratic regime (e.g., unless court order was provided).

Public key infrastructures will probably result in a nightmare for private and corporate users worldwide with substantial costs to be paid by users, and misuse and abuse by unauthorized parties being an ever-possible threat. Current discussions in France about legislative efforts for public encryption do not suggest a practical solution to this problem (e.g., http://www.freenix.fr/netizen/chiffre/advicece.html).

SUMMARY AND CONCLUSION

This chapter outlines how deregulation of markets and regulation affecting the Internet and electronic commerce (e.g., digital signature) has evolved in industrialized and developing countries. It is important that although North American and European users and users in some Asian/Pacific countries (e.g., Japan and New Zealand) may benefit from deregulation in telecommunication and cable services, most people around the world are still awaiting electricity and/or telephone access from their home. Nevertheless, this chapter also illustrates how national sovereignty may be replaced with international sovereignty in order to develop legislation enabling the Internet to really be used to its fullest capacity. Particularly, the struggle by EU countries to use supranational forums for addressing these issues and identifying the measures to be undertaken based on shared moral and ethical frameworks adhered to by people living in member states (see chap. 5) indicates that we still have far to go.

At this stage, a lack of harmonization between regulations across states (e.g., United States) and countries (e.g., EU) hamper electronic commerce. In turn, a U.S. employee sending an encrypted message from France to home may violate French law unless the encryption software is registered with the appropriate French agency. However, if the employee waits and sends the message from Frankfurt Airport, no local laws are being violated. Similarly, certain U.S. encryption technologies are provided to some people (e.g., U.S. and Canadian residents) whereas others are being excluded (e.g., EU) due to political and national security concerns. Moreover, some technologies such as Tempest (see Appendix A for an explanation) are not even available to the public. In turn, how can a citizen protect his or her right for privacy if one is not even being aware of having one's privacy violated by some security agency?

In summary, this chapter illustrates that although many countries have tried to deal with the rapid diffusion of Internet technology into households and businesses, many are just beginning to establish the legal frameworks necessary to handle consumer and other concerns. Moreover, in some countries, Internet access may still take years to come until the necessary infrastructure has been established such as providing drinking water and electricity to many communities. Nevertheless, the deregulation of telecommunication markets in industrialized countries and the beginning of satellite communication (e.g., Iridium) becoming a competitive option for some has helped many private users to gain relatively inexpensive access to the Internet. Nevertheless, some vast differences still exist and these are addressed in chapter 3.

3 Personal and Organizational Use of the Internet: Economic and Access Issues

OVERVIEW

Access to the Internet for business, educational institutions, and private citizens is a challenge. One reason is that only about 4% of the people in the developing world have access to a phone. Even in developed countries, access charges (e.g., to connect to the Internet through cable and Internet charges itself) may slow down the diffusion of Internet usage by business and private citizens. For organizations, labor, hardware, and maintenance costs for an in-house or a virtual server for storing Internet material are also a matter of consideration. These and other possibly negative and positive developments are outlined and discussed (cf. Appendix C).

Interdependence, social complexity, diffusion, and structure of how people take advantage of CMC, and the Internet in particular, are important issues as chapter 1 has outlined. Moreover, market deregulation in telecommunication and cable TV service providers also influences the pricing one pays to gain Internet access via telecommunication, cable, and in some cases, even electricity lines as discussed in chapter 2. Moreover, regulatory developments across borders or lack thereof in such areas as encryption and digital signatures influence the opportunities, costs, and attained safety or security for electronic commerce. Accordingly, institutional norms about communication and procedures about doing electronic commerce are affected by both deregulation and regulation.

Economic issues linked to CMC and the Internet are also of considerable importance when it comes to future developments and institutional progress. Here, we have to distinguish between private users and organizational ones. In both instances, costs for hardware, software, access, and Internet use itself is of importance. Important differences exist between these two groups of users. For instance, organizational users have to determine if hardware, server, software, and communication as well as labor costs warrant an in-house or a virtual server instead. This chapter outlines economic and access issues for private and organizational users of the Internet.

INTERNET ECONOMICS: WHAT DOES IT COST?

Today, most people link to the Internet by using a modem to dial a phone number that then connects them to a modem on the other side. Once this connection is established, the user can gain access to the Internet. Alternatives to the telephone such as using a cable modem or electricity wiring for connecting to an Internet gateway are increasingly used in industrialized countries. We concentrate on phone access costs in this section because most people and organizations still use this technology to connect to the Internet. Moreover, in developing countries, even phone access is still the privilege of a small minority, whereas cable service rarely, if ever, exists beyond the capital.

Getting Connected to the Internet: Current Cost Impediments to More Extensive Use of the Internet

There are several ways to get the physical connection to the Internet and the one you choose depends mainly on the volume of information flow, or traffic, that you will have. A *dial-up account* with an Internet service provider is a slow speed connection that uses a telephone line (usually up to 57.6 Kbps). In February 1997, U.S. Robotics began shipping its proprietary x2 modem technology that allows Internet connection at 56 kilobits per second, which at the time was about twice as fast over current phone lines as other modems. Each connection requires a separate account, and all information transactions are done while you are connected. A *SLIP/PPP account* gives you the option of having a dial-up or a full-time dedicated connection with a telephone line. It is more user-friendly, especially for unsophisticated users, as it allows access with Windows and many tasks can be performed simultaneously. A *leased-line connection* is the best option for organizations who have large numbers of users and large amounts of information traffic. It provides fast access that is available around the clock. However, the faster the connection, the greater the cost.

In most developed countries, a user can get access to the Internet by communicating with another modem or computer using the telephone, cable services, and in some instances, power lines or other technology. The most obvious ones are naturally the telephone and the cable. However, since 1997–1998 electric utilities in Denmark, England, and Germany have been trying to iron out small glitches in the technology, thereby permitting these firms to deliver Internet and telephone services using their network of power lines. In 1997, an English utility was starting to sell this service to its subscribers.

Speed of Information Transfer.   In some countries, customers can choose the type of line their telephone company provides. The telephone line will determine the costs, and also the type of information that can be accessed and the speed with which large files can be downloaded to computers. In the case of Integrated Services Digital Network (ISDN) and Asymmetrical Digital Subscriber Line (ADSL) access, both the private and corporate user as well as the Internet access provider require the same technology to achieve the highest data exchange rate. ISDN provides greater bandwidth than normal telephone service. ADSL is a new technology that was introduced in 1996 in some U.S. markets, and during 1998, in Germany. In turn, this has resulted in pushing ISDN prices down. ADSL allows transmission speeds of up to 10 times the speed of ISDN, but in one direction only. Accordingly, it is ideal for users who want to either download or upload vast amounts of data.

Cable companies can offer customers greater bandwidth than either ISDN or ADSL. Cable companies providing Internet services in Canada, Denmark, Sweden, the United States, and England, to mention a few countries, are increasing the pressure on telecommunication carriers to improve services while reducing price. Cable access is about 200 times faster than normal telephone access and about 80 times faster than ISDN service. For those who download videos and/or use the Internet for long distance telephone calls, large bandwidths are necessary to allow near real-time images and sounds. However, cable service or an ADSL modem permitting fast downloads offer clients viable alternatives, hopefully, in turn, resulting in competition between firms, thereby enabling clients to choose the most appropriate service at a favorable price.

PRIVATE USERS

In industrialized countries, anywhere from 10 to 50% or more of all households (e.g., more than 50% in Denmark and the United States since late 1998 and rising) do have a personal computer at home. Naturally, this does not mean Internet access but the most important hardware required is available. This percentage is likely to rise further for two reasons:

1. Low end computers are now available for under $1000 (U.S.) enabling users to surf the Internet.

2. More and more, European employers are selling their older machines for a song to employees to avoid paying high environmental charges to have such technology disposed of legally and properly (e.g., paying $300 or more for the environmentally appropriate disposal, versus selling a computer for $75 to an employee).

The previous discussion indicates that in industrialized countries, obtaining the necessary hardware required to access the Internet is no longer a financial burden. Nevertheless, variable costs incurred by surfing on the Internet may vary greatly across countries.

Communication Costs

For private users in North America, it is a straightforward matter to obtain unlimited telephone service for about $15.00 to $30.00. The story is much different in most other countries. For instance, consumers in Switzerland pay a set fee to have telephone service (approximately $18.00 per month), and there are even charges for local calls. The time charges differ based on the calling time, duration of the call, and the calling distance. Table 3.1 provides the reader with some cost comparisons and indicates that in some countries, Internet access using telecommunication means can be expensive, and alternatives are limited (see Table 3.1). These issues are further discussed later.

As outlined in chapter 2, deregulation of various markets has resulted in cases whereby cable companies may compete fiercely with telecommunication firms for Internet customers (e.g., Denmark). Moreover, in some instances, logging on to the Internet no longer requires a pay-per-minute fee or a fee to a telecommunication or cable provider if they have offered such access for free. For instance, in The Netherlands, people with cable TV are offered Internet access as a free add-on service to basic cable. The average monthly fees paid according to metered Internet service are lower in comparison to other technology (e.g., paying the phone company for connecting with a phone modem as well as the Internet provider for use of the latter).

Nevertheless, Table 3.1 indicates that if a person does pay for local phone or cable charges for connecting to the Internet, these are big enough to cause some pain to the private user in some markets. Again, the charges

TABLE 3.1

Accessing the Internet: Monthly Charges for Telephone, ISDN, ADSL, or Cable (Using 1997 UBS Purchasing Parity Ratios[a])

Country	Telephone	Telephone Private Users ISDN[b]	ADSL	Cable
Canada	Unlimited local calls, U.S. $12.52–16.69	U.S. $66.76	U.S. $25.04–37.55 available as of 1997 in most Canadian markets	U.S. $14.60–16.69c
U.S.	Unlimited local calls U.S. $15.00–25.00	Unlimited use U.S. $249.00d, or U.S. $31.00 for 20 hours/month	U.S. $35.00–40.00 available as of 1997 in most U.S. markets	U.S. $17.00–35.00
Germany	U.S. $18.36f (DM 24.60)	U.S. $34.22 (DM 46)g	not available as of 1998	not available as of 1998
South Africa	U.S. $20.86 @ 50.00)h	U.S. $62.59 @ 150.00)	not available as of 1998	not available as of 1998

Note. An earlier version of this table appeared in Gattiker, Janz, & Schollmeyer (1996); some changes and expansions have been made here. All figures are based on 1998.

[a]Purchasing Parity: Allows the comparison across countries of costs using a basket of goods to arrive at purchasing parity that is often quite different from the official or black market exchange rate. The Union Bank of Switzerland produces such a comparison for numerous cities around the world in order to help organizations to remunerate expatriate managers appropriately and fairly.

[b] ISDN costs about U.S. $66.76 for 128 kbps single-user dial-up service for the user; however, the Internet provider pays Cdn. $80.00 for each ISDN access. Hence, the user may have to bear some of these additional costs from the Internet provider as well, bringing the ISDN costs for the Internet to over U.S. $83.45 per month.

[c] In spring, 1996 Canadian cable operators announced that by fall of 1997, Internet access would be available to all subscribers. The service would require the user to either purchase a standard cable box or rent it. Hence, limited additional costs will be accrued by subscribers for obtaining Internet access in addition to what is already paid for basic cable services. By the end of 1997, service was about ready to be launched in a few selected markets in Southern Ontario and available elsewhere by mid-1998.

[d] Bell Atlantic (a U.S. telecommunications carrier) lowered its ISDN rates by 15–86% in spring of 1996 for subscribers in New Jersey, Delaware, Pennsylvania, and Virginia. Bell Atlantic offers a full-time connection for U.S. $249.00. Additional price cuts are expected with increased competition from cable operators and ADSL services. Moreover, since February, 1997, U.S. Robotics has shipped its proprietary x2 modem technology that permits Internet connections at 56 kilobits per second, roughly half as fast as ISDN but without the costs.

[e] Field trials started in 1996 and U.S. cable operators intend to offer local telephone services as well as Internet access to their subscribers. Operators are lobbying the government to permit subscribers to keep their telephone numbers when switching telephone services from a telecommunications carrier to a cable operator.

[f] Additional costs are incurred for each call on a per time-unit basis according to peak or off-peak rates.

[g] For 10 hours = DM 75.00, for 20 hours = DM 90.00.

[h] Additional costs are incurred for set-up charges for each call as well as time used, the latter changes according to peak and off-peak times.

are adjusted for purchasing parity in developing countries such as South Africa. Moreover, comparing these prices according to living standards makes Internet access from home for people in developing countries a near impossibility.

Internet Access Costs

Although communication costs must be considered before one is able to surf on the Internet or send electronic mail, one needs a way to connect to the Internet. Once telephone or cable access is secured, a user has to decide which Internet service provider to select and, most importantly, needs to determine how much access charges are being incurred for surfing on the Internet. Naturally, Internet access can also be provided by the telephone or cable company and it often is (e.g., German Telekom).

In 1996, AT&T decided to offer unlimited Internet access to its long distance telecommunication customers (80 million in the United States alone) for under U.S. $20.00. In the United States, America Online and CompuServe offered unlimited Internet access for under U.S. $20.00 even before AT&T. This offer was made based on their usage figures showing that their average clients used the Internet for under 7 hours per month. Nevertheless, after providing its U.S. and Canadian customers with unlimited access for a fixed fee in late 1996, America Online (AOL) discovered a surge in usage from about 7 hours up to about 30 hours per month for its average user. But by January 1997, this rapid increase in usage had taxed AOL's system so much that only about 3% of all users were able to be online at any one time. To hold off lawsuits by angry customers being unable to connect to its service, AOL promised to improve the situation to permit about 5% to be online at any one time by installing more access modems and better technology. Unfortunately, this still puts AOL at the trailing edge of the industry's average, which usually enables 5 to 10% of all users to access the Internet at the same time (Meeks, 1997). In comparison, in industrialized countries, more than 10% of telephone subscribers can place a call at one time without overloading the system.

Although AOL has raised prices in early 1998, it is still hard to generate much profit from offering unlimited surfing for below U.S. $30.00 per month if the average user spends more than 30 hours on the Internet. Nevertheless, this illustrates that thanks to economies of scale and market competition, for a reasonable monthly fee unlimited Internet access is no longer an issue in the United States (see Table 3.2). Similar developments have occurred in 1996 in Canada, albeit remote communities in scarcely populated areas of Canada's north still have difficulty in having much of a choice between Internet providers.

This discussion indicates that for North American users, unlimited Internet access including phone or cable charges is probably not much more per month than going out for dinner as a couple and taking in a movie afterwards. Hence, for the individual user, it is simply an issue of either giving the Internet priority and thus paying the fees out of one's disposable income or going out for dinner and a movie once more than one would otherwise. It definitely does not represent financial hardship. Hence, Freenets or Savenets are rarely needed to obtain economies of sale in North America and using public institutions such as libraries for Internet access is left to a minority only (e.g., not having Internet access due to economic hardship or unemployment or not having a computer at home).

In Europe, the picture is relatively different because large providers such as AOL Germany do not offer unlimited services. Additionally, only in few markets is accessing the Internet via the cable fixed, that is, only Internet charges are incurred but no cable communication charges (e.g., the Netherlands).

In a developing country such as South Africa or Brazil, its obvious that a small minority, most likely in urban centers has the financial resources to secure private access to the Internet. For the majority, however, access through public institutions providing hardware and having leased lines to the Internet such as schools and public libraries is probably the only viable option for many people in cities and rural areas (see also chapter 2). As the following section shows, however, organizational users have different concerns about Internet access than do private ones.

ORGANIZATIONAL USERS

In organizations, nearly everybody who works in an office and even on a production floor uses some type of information technology including personal computers to do their job. In the following sections, I focus on small and medium-sized enterprises (SMEs) for whom cost issues are probably more of a concern than they are for large organizations.

Hardware, Software, and Labor Costs

A server may cost a firm anywhere from $10,000 up to $40,000 or more whereas traditional costs, such as software and tape backup systems, may add a few thousand dollars more to the price tag (see Table 3.3). Highest are

TABLE 3.2

Comparing Internet Access Costs Across Countries

	Basic Fees	WWW-Fees per Hour	Telecom Charges Low Tariff During Evening/Weekends for 5- to 6-minute call	Average Charges for 10 Hours Online and About 60 Messages in and/or Outgoing per Month	Average Monthly Charges in U.S. $ Using the UBS Purchasing Parity Ratios (Approximate Numbers Only)
Switzerland					
EuNet	Sfr. 25.00	Sfr. 25.00 unlimited access	City = Sfr. 0.30 > 10 km and more = Sfr. 0.90	Sfr. 48.00 local[a] Sfr. 89.00 if trunk call[b] charges apply	U.S. $15.02–45.90
South Africa					
Aztec	R 50.00	R 2.00 per hour or R 95.00 for unlimited use	R 0.25	R 95.00	U.S. $40.89
Bulletin Board	R 10.00–R 20.00/month	R 5.00–R 10.00 per month regardless of amount of use	R 0.25	R 55.00	U.S. $25.04
CIRC	R 0.00	R 0.00	R 0.00 if access is done from CIRC terminal. Local telephone call charges apply if system is accessed from home	R 0.00 if access is done from CIRC terminal	U.S. $0.00

Canada					
Victoria Free-Net	Cdn. $0.00	Cdn. $0.00	Cdn. $0.00	Cdn. $0.00	U.S. $0.00
Cadvision, Calgary	Cdn. $150 per annum	Cdn. $0.00	Cdn. $0.00	Cdn. $0.00	U.S. $11.68–13.35
Internet Services Lethbridge	Cdn. $30 ($30 set-up fee)	90 hrs of connection time included in basic fee, $2 for each additional hour, users are currently not charged for exceeding 90 hours	Cdn. $0.00	Cdn. $0.00	U.S. $28.37
Australia					
Australian Open Net	AU $0.00	AU $0.00	AU $0.23 regardless of length of call	AU $64.60	U.S. $79.28

Note. An earlier version of this table appeared in Gattiker, Janz, & Schollmeyer (1996); additions and changes have been made. The data represent estimates based on information obtained from various sources during 1998. The information provided here indicates that users in Canada have the opportunity to obtain full-fledged access to the Internet at the lowest cost considering their high standard of living. Similar numbers would materialize for U.S. users. Switzerland and other European countries in general, due to a lack of telecommunication deregulation having yet lowered local phone charges substantially reduces downward pressure on pricing, are being charged a higher amount in comparison to other countries. Most importantly, the table shows that people living in more remote areas of a country (e.g., Canada and Australia) pay substantially higher Internet access charges to Internet service and telecommunication providers.

[a]Basic fee + 10 hrs of WWW access, telephone charges to access and check mail 30 times a month.

[b]If the person lives beyond 10 km of the access point, the telephone charges increase from Sfr. 23.00 to Sfr. 64.40 [(30 x Sfr. 0.70 to check mail = Sfr. 21.00) + (14 x 30 Min sessions on the WWW at Sfr. 4.60 per 30 Min.)], an increase of Sfr. 41.40 per month assuming that the person does not get disconnected a few times a month, which generally occurs, increasing charges again.

probably not the costs for purchasing a server but, instead, the expenses paid for maintaining, upgrading, and managing the server. This in turn requires the firm either to outsource such work or to have qualified personnel on hand to assure that the system is running smoothly 24 hours a day for 365 days a year. Here, costs may be very high for some countries where work after certain hours or during public holidays and weekends has to be compensated with a special bonus as agreed to by unions and employers. Here, the firm faces two issues; direct and indirect labor costs for qualified personnel; and rates to be paid for working overtime or during nights and/or weekends or on public holidays.

In a developing country such as India, qualified technicians may be hired costing a firm less than $1,000 a month including indirect labor costs and vacation pay and, most importantly, shift work and also working on public holidays. This may be one reason why companies such as Siemens have opened shop in Bangalore, India for doing much of the programming and maintenance work for software and hardware used by the firm worldwide. As expensive can be having to pay according to local labor laws and/or union contracts that may stipulate stiff surcharges for employee pay if working after regular business hours and on weekends or public holidays is required.

Based on direct and indirect labor costs as well as on regulations concerning shift and public holiday pay in most European countries, unfortunately, high labor costs will make it impossible for an SME to meet the 24 hour at 365 days performance goal for its server(s). At minimum, two specialists are needed by the firm, more likely three to do a good job and to assure coverage even during annual holidays. Some firms may outsource the management of a server to another organization while having the server still on their premises. But even the firm managing the server on behalf of its clients may not have 24 hour service but, instead, emergency service on standby. In turn, if a server is not online, the technician being on call might be automatically informed with a beeper and may be required to do a service call in the middle of the night in the hope of fixing the problem quickly.

As Table 3.3 suggests, a virtual server might be a viable option for the firm to save labor costs, while hoping to maintain a 24 hour for 365 days performance standard. However, unless the virtual server provider is able to have technicians working 24 hours a day, this performance standard is unlikely to be met. In Europe, most providers are unable to guarantee such service due to high labor costs.

TABLE 3.3

Cost Comparison Between In-House Versus Virtual Server

Costs for an In-House Solution (in US$)

	1 Time Fee	Annual Fee
Technical:		
T1 + provider	2,600	12,000
Hardware	10,000	2,000
Software	10,000	Updates
Web Design	?	?
Use:		
Personnel	-	80,000
Domain	-	50
Total	22,600	94,050

Costs for a Virtual Server (in US$)

	1 Time Fee	Annual Fee
Installation:		
Setup (T3)	100	-
Software	150	-
Web design	?	?
Use:		
Personnel	-	-
Virtual server	-	895
Domain	-	50
Total	250	945

Note. By 1998, it was possible to purchase for about U.S. $1,250 a server (e.g., Apache) offering a server with HTTP1.1. CGI, SMPT, IMAP4, POP3, and FTP services. At the time such a machine could manage 140,000 e-mail messages, 50,000 data transfers, and 250,000 http requests per day. It had a 150 MHz Mips-processor, 16 MByte main memory and a 2.1 GByte hard-drive. Such small servers make it feasible for private users as well as small firms to have their own server in-house if permanent Internet access can be secured at a relatively inexpensive price.

In some countries, telecommunication or cable access providers to the Internet backbone do, however, charge the user based on the volume of traffic. Accordingly, in contrast to the United States, many European users will not enjoy fixed access charges for leasing a T1 or T3 line in case of an in-house solution. Some virtual server hosting companies simply absorb these costs and offer their clients a fixed price assuming that once high traffic volume occurs, organizations will want to have their in-house server anyway.

Communication and Access Costs

As the U.S. based pricing in Table 3.3 used during 1998 shows, managing one's own server can be a costly affair. Moreover, if one is located in a market with limited competition, telecommunication or cable charges incurred by a firm are likely to be higher than in the United States or England. Again, this represents an obstacle for an SME to establish a competitive Web presence.

A U.S. or Canadian firm may be able to lease a permanent ISDN or T1 line for $250 and/or $1,000 respectively compared to $1,000 or $10,000 in a country such as Germany where Telekom is still the dominant provider of such services. Again, such costs can be simply prohibitive for an SME firm.

Accordingly, to avoid high communication charges, (e.g., telephone of TV cable) many medium-sized firms use virtual servers or place their server into the premises of their U.S. or Canadian subsidiary if this is a possibility. In turn, costs for permanent Internet access as well as for managing labor and maintaining the server can be saved, although most content on the server will be in English as well, the dominant Internet language.

Virtual Versus In-House Server

Based on the earlier information and on Table 3.3, it seems obvious that an SME might consider leasing space on a virtual saver that could be located anywhere around the world. In the case of a virtual server, access costs or telecommunication charges incurred by the provider of the virtual server are shared by many users. Hence, the virtual server option might be a viable alternative to save telecommunication as well as labor and hardware costs by a firm.

Because of low communication charges as well as more flexibility and fewer restrictions in the labor market, U.S. virtual server providers have become a force offering their services to clients worldwide. Their competitive edge over competitors is currently hard to overcome, unless better service and more customer-focused products are offered by a provider in Europe such as Denmark. Nevertheless, some firms may prefer to have their server physically nearby and cultural differences (see also chap. 4) may facilitate the working with a local provider of a virtual server, possibly offsetting the higher costs to some degree if the virtual server is in Europe.

Today, many organizational users pay American providers of virtual servers a monthly fee depending on their server needs. However, only the future will tell if further deregulation of telecommunication and cable services as well as of labor markets will result in cost levels making virtual servers or even in-house servers a viable alternative for European firms instead of having to go overseas. Changes in hardware making a small server cost little more than U.S. $1,300 (see also Table 3.3. footnote) make telecommunication and/or cable charges for Internet access more and more the sticking point for SMEs trying to run their own server. Private users often are able to receive free space on a server somewhere, most likely also located in North America. Here the hope is that large groups of users of free Web space share similar interests, entertainment, and personal finances. Such virtual communities tend to become self-supporting if their membership moves beyond 5,000 people sharing a particular interest (e.g., home repairs and cooking for consumers, financing and selling or purchasing components for businesses). In turn, such a site will attract enough advertisers to pay for the free server space offered to "community members" to have some of their own information on the Web (i.e., space provided for storage is not unlimited; see also chaps. 7 & 8).

The previous discussion and data in Tables 3.2 and 3.3 suggest that although today a virtual server in the United States might be a viable option, tomorrow, further deregulation in telecommunication markets and lower hardware costs for a server as well as software requirements may again change the picture that permit an SME or even a private person to have one's server in-house.

Externalizing of Costs by Private Users

Even if the firm has decided about reducing its server costs as much as possible by choosing a mixture of the alternatives outlined earlier and in Table 3.3, externalizing of Internet costs by employees still has to be considered as an important issue. Economists suggest that certain actions by individuals result in positive and negative externalities that affect others (e.g., DeSerpa, 1994). Externalities can also be conceived of as unpriced byproducts of (inputs to) a person's behavior (Wijkander, 1985). Negative externalities may require some action by the government to tax certain users (e.g., polluters) whereas in a health insurance scheme, an attempt might be made for reducing free riding on part of some consumers or clients (Wirl, 1994).

How does this apply to Web usage? We have no empirical evidence to suggest that externalizing of costs leads to different behavior and attitudes on the Internet. However, we do have data indicating that telephone pricing does affect the originating of traffic, that is, the product is sensitive to pricing (e.g., Donzé, 1993). An unpriced byproduct for an employee with Web access at work is using the Web after working hours for private use. This reduces his or her costs in several ways if the employee has metered telecommunication service from home (fee per call and time); and Web access from home to a provider is metered as well (e.g., 3 hours free with monthly fee, for each additional minute, Internet access provider charges users).

Hence, those costs are likely to be externalized (i.e., saved) by the employee if he or she is using the employer's facilities after regular working hours.

If the employee can connect to the Internet account one might have at work from one's home access charges can be saved in one of two ways:

1. One may not have a service for private Internet use, thereby saving these monthly expenses (see Table 3.1), or
2. Telecommunication costs can be reduced by using a call-back feature, whereby one calls a number and hangs up and is called back by the employer's computer, hence the employer picks up the tab.

In cases where the telephone or cable charges are fixed per month regardless of how often or for how many minutes one uses the phone for making local calls (e.g., United States), call back will not be used.

For reducing externalizing of costs by employees, the organization has two possibilities:

1. Restrict Internet access and use to work-related applications or needs only as done by the U.S. federal government. Violations could result in one's dismissal; or
2. Permit the use of the Internet from home to reduce its use for private purposes at work.

Restricting Internet access might be a viable option. Nonetheless, enforcement of U.S. government guidelines has been used primarily in a few cases such as individuals downloading pornographic pictures (e.g., Lawrence Livermore Laboratory) or doing online shopping (U.S. Army Europe) for personal use. However, restricting or maybe even preventing employees from using the Internet at work for private purposes is often difficult to manage. Worse is if it might not even be in the best interest of the organization. For instance, online shopping for U.S. Army personnel stationed in Europe might help in easing the "pain" in being stationed abroad. Moreover, some TV stations in North America report that the heaviest traffic is during lunchtime, when viewers check out information about their favorite show and its characters, plus sports and weather information. If such activity occurs not during work hours but during lunchtime instead, the cost for the employer is minimal if not lower than having the employee make private phone calls from work. Most importantly, the employee might acquire additional Internet skills if not even helpful information for work when surfing on the net (see also chap. 8).

Naturally, use of Internet facilities during work hours is as undesirable as employees making their weekend arrangements during those times but using the firm's phone system instead. Unfortunately, unless companies have clear policies such as permitting use for private purposes during lunch break and after working hours (cf. Appendix F), there may be some potential conflicts in the making. Finally, permitting the use of the Internet by employees at agreed upon times or hours represents a desirable benefit from the employee's perspective, while costing the firm little, if anything. Fixed costs, such as leasing a T1 line, remain the same whereas the server or other hardware should be running 24 hours anyway. Moreover, in some countries one can often enter the work facility during weekends only if one is scheduled to work overtime; if this is not the case, access is denied due to liability (e.g., person is not insured when having an accident at work) and labor laws (e.g., union agreement). All these concerns fall by the wayside if the employee accesses e-mail from home and writes a reply to a customer before coming back to work after a long weekend.

Accordingly, although externalizing of costs incurred through private Internet use at the employer's expense is unlikely to be eliminated, sensible policies may benefit all parties, whereas employees answering e-mail or finding important work-related information during off work hours do not have to be paid wages for these hours either!

SUMMARY AND CONCLUSION

This chapter illustrates that private and organizational users in developing and industrialized countries have different concerns. Cost factors are probably a far greater impediment to home use of the Internet than to commercial use. Nevertheless, high costs for leased lines in some markets makes it difficult for SMEs to lease high capacity lines, which, in turn, reduces access speed for customers. The latter may not necessarily have the patience to wait and incur additional charges if their phone and Internet access is metered. In turn, this could result in many lost sales by users moving on to the next site.

The cost structure outlined here also indicates that access charges in developing countries may be such that it will take a long time until a large group of people will be able to afford the necessary hardware and monthly expenses permitting Internet access from home. For organizational users, however, lower labor costs to service and maintain an in-house server 24 hours for 365 days a year could be a competitive advantage whereas charges for leased permanent Internet access lines may eliminate much of this cost advantage again. However, the many firms in North America offering reliable and economical virtual servers to firms around the world would suggest that the combination of low charges incurred for permanent access lines to the Internet and reasonable labor costs add up to a competitive advantage. In turn, this has resulted in a new virtual server industry in the United States competing for clients worldwide.

The economic and access issues outlined here also influences who the major social actors are and their capacities for actions affecting institutionalization. For instance, in some countries, private access from home is limited whereas in others, it's the norm. Moreover, the number of private users will affect content, norms, rules, and interaction on the Internet. It seems quite safe to assume that where a large percentage of households access the Internet, its content, services, and electronic business opportunities will differ then from countries where the majority of users are organizations. In the latter case, business-to-business e-commerce may be the only interesting game in town, whereas in markets where households surf extensively from home, business-to-consumer contact and sales via the Internet may become an interesting business opportunity for many SMEs.

For private users in North America, Web surfing or other Internet activities have become a good that is nonexclusive. This means that for a fixed monthly fee paid to the Internet service provider (or at zero cost to oneself if one uses one's workplace as an access ramp from home) or to the phone company for making unlimited use of telecommunication facilities locally (in some cases through the cable company), the private user in the United States or Canada can increase benefit or consumption of the Internet/Web without increasing one's contribution to the scheme. Accordingly, it is easy for a North American user to leave his TV on while surfing the Web simultaneously or while being online constantly during most evenings. In turn, this requires some Internet providers to restrict Internet access for private users during evenings, otherwise additional costs are incurred by the user. This chapter illustrates that in markets where users can minimize or fix their costs, Internet use as measured by time will likely be higher than in countries where charges are metered and variable (i.e., more expensive during day hours).

II How Culture, Attitudes, and Beliefs Influence the Use of the Internet

Part I of this book focused primarily on hard facts, that is, how the Internet functions, regulatory matters, and Internet policies as well as economic and access issues. In part II, the focus is more on what could be called soft issues. For instance, as chapter 4 outlines, infrastructure (e.g., telecommunication) is the objective side of culture as is the regulatory framework. However, both can change over time. In addition to the objective side of culture, there is also the subjective one including people's beliefs, attitudes, and behaviors (see Fig. 4.1) possibly being moderated by norms, rituals, and values.

Part II focuses on better illustrating how cultural similarities and differences in conjunction with infrastructure and policies as outlined in part I may influence the organizational and private use of the Internet. Chapter 4 discusses a cultural framework and outlines these issues in how they may affect communication and CMC in particular by comparing Internet communication to television and other media. Chapter 5 brings this a step further and outlines the moral and ethical issues we face with CMC and how, based on cultural similarities or differences, we may communicate in various ways. This chapter illustrates how difficult it is to develop some institutional characteristics for the Internet because even though geographical boundaries may no longer be an obstacle to people communicating with each other from far away places, misunderstandings due to culture are plenty to be found. In addition, regulatory frameworks taking these cultural differences into consideration are very difficult to develop.

The final chapter in this section is chapter 6, which focuses on how ethical and moral understanding across nations may result in differences in how

Internet users interpret situations, behavior, and make sense out of what is happening in cyberspace. Hence, privacy and using customer information gathered on the Internet for marketing may be perceived differently by people with different cultural backgrounds. Although a code of conduct or ethics regarding one's work with computers and the Internet could be helpful, these codes differ even between professional societies within a country but across disciplines (e.g., computing and medicine) as well as between countries (e.g., two computer science associations). This illustrates that regulatory efforts for having Internet users follow one ethical framework while using similar moral standards (e.g., what is decent versus indecent information?) across countries may be futile.

4 Cultural and Cross-National Issues

OVERVIEW

To what extent is there a new culture evolving in cyberspace, with its own values, beliefs, and norms adhered to by its users? Subgroups, such as hackers and cyberpunks, may adhere to different norms than other users. Cyberspace is part of the postmodern epoch, where information and knowledge replace capital and labor. However, whereas geographical boundaries may be overcome, cultural differences could hinder the Internet from becoming an institution with shared values, norms, and rituals. The interrelationship of these issues with cyberspace and culture are outlined and opportunities and risks offered by these developments are discussed (cf. Appendix B).

Part I and chapters 1 through 3 set the stage in so far as they outline how the Internet has offered new opportunities for communicating around the globe. Governments have tried to set policy and regulation where necessary, which in turn has affected the economics of Internet useage by consumers and organizations. Often, however, people are unaware of how cultural differences may affect Internet users. Our values, beliefs, interests, and objectives guide which customs, norms, and rituals we adhere to and perceive as morally acceptable. Accordingly, culture provides the foundation for our understanding of justice and, ultimately, of the law as well as how we interpret and administer our laws (e.g., regulatory concerns). In turn, culture also influences our cyberspace choices, practices, the design of technical systems (e.g., video game entertainment), and the possible development of institutional characteristics for the Internet.

79

This chapter focuses on cultural and national differences as well as on similarities and how they might influence CMC and the use of the Internet by citizens and organizations. First, a discussion of how culture can be defined is given and television and cyberspace entertainment are used to illustrate similarities and differences between countries. Finally, cyberspace as a cultural genre is discussed and possible future developments are outlined.

THE MEANING OF CULTURE

Any society may be considered as having a variety of cultural "themes," rather than a single culture. Cultural themes are composed of various interpretations and heterodoxies of the core culture, in addition to any incursions that may have developed around the core, such as the introduction of a new ethnic group. Cultural diversity in countries has been increasing due to the internationalization of business; the workforce has become more variegated due to the entry of guest workers, immigrants, and refugees. Cross-national studies about individual and organizational phenomena are concerned with the systematic study of the behavior and experience of cyberspace participants in different cultures. A brief discussion of the most pertinent cultural issues follows.

Most anthropological studies contain one or more of the following cultural derivatives; *symbol* (including language, architecture, and artifacts), *myth, ideational systems* (including ideology), and *ritual*. Whereas anthropologists and sociologists continue to debate the correct usage and meaning of these concepts, most studies treat them as motivational factors for individuals and groups (Silverman, 1970). Psychologists tend to follow Triandis and Vassiliou's (1972) model whereby a distinction between *subjective and objective culture* is made. Subjective culture is defined as a group's characteristic way of perceiving its social environment. For example, office workers could differ in their attitudes toward computers based on demographic characteristics such as gender (e.g., Gattiker, Gutek & Berger, 1988).

In the context of this book, it is assumed that culture represents both a stable and an individual/environmental dimension (objective/subjective continuum using Triandis' terminology; see Gattiker & Willoughby, 1993 for an extensive discussion of this issue).

As Fig. 4.1 reflects, the x axis (horizontal) is a continuum that ranges from the micro focus (i.e., the individual) to the macro perspective (i.e., the environment) of a culture; and the y axis (vertical) represents the level of stability of the culture ranging from low stability [e.g., approximate subjective (opinions)] to high stability [e.g., innate subjective (cognitive style)].

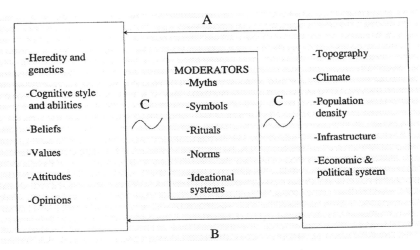

FIG. 4.1. Relationship between the micro–macro continuum and the degree of cultural stability.

The micro dimension is represented by the cyberspace user whose attitudes and opinions are likely to change frequently during his or her lifetime (low cultural stability), that is, what is "cool and in" today may be "out" tomorrow. The macro side represents the natural and humanmade environment. Whereas topography may remain stable over thousands of years, a country's telecommunication infrastructure can change rapily after deregulation (e.g., 1997 when the European Union deregulated telecommunications, although the United Kingdom had opened its telecommunication market to competition in 1991). Moreover, if cable operators, railways, and electrical utilities are also allowed to offer telecommunication services to the public, further changes will occur.

Arrow A symbolizes the influence of the natural environment on the individual. For instance, a change in climate may lead to the survival of only those individuals whose genetic makeup, as the result of favorable mutations over generations, has adapted them for survival.

Arrow B symbolizes the bidirectional relationship between the approximate factors, such as the individual's beliefs, values, attitudes, and opinions about the Information Highway, and the humanmade environment, as represented by the infrastructure and the economic, legal, and political systems of a country.

Wave C illustrates the intermediary effect of cultural moderators on the approximate individual factors and the humanmade environment. Accordingly, symbols about Netiquette when sending e-mail may influence how a society responds to a person's behavior whereas it might be frowned on to send advertising for products via the Internet. Hence, when a couple of lawyers sent advertisements about their book dealing with cyberspace issues via the networks in 1994, people got angry and reinforced the rules by sending them thousands of messages until their mailbox had to be shut down (see also Table 7.4).

The literature does not support the view that cultural moderators and the natural environment affect innate individual factors. However, this does not mean that a certain topography and climate may not foster certain myths and symbols. For instance, the Inuit language contains over 30 words describing snow, and Inuit fairytales likewise reflect the importance of snow and ice.

The left rectangle represents the individual dimension of Fig. 4.1. Hence, its location is to the left (x axis = micro focus), and the stability of these factors decreases from heredity down to opinions (y axis). For instance, public opinion polls show that the electorate frequently changes support for the government in power. In contrast, people's beliefs are relatively stable and resistant to change (e.g., Rokeach, 1980). This illustrates that when we try to comprehend culture from a micro perspective, we must accept that an individual's opinions are less stable than his or her beliefs. Moreover, although we can measure genetic factors such as eyesight and reproductive behavior, it is far more difficult to comprehensively assess opinions. Whereas opinions are *approximate*, heredity and genetic factors are *innate*; they occur as a result of genetic mutations over generations (Plomin & Rende, 1991).

The right rectangle in Fig. 4.1 graphically illustrates the environment that is, at the top, *natural*, and at the bottom, *human*made. Similarly, although the natural parts of the environment, such as topography and climate, are stable over generations and centuries until the next natural disaster, population density or infrastructure, including roads and the Internet, result in the implementation of human actions and policies. To illustrate, low Internet access costs in North America enables users to spend their spare time communicating on chat lines, reading newspapers online, and obtaining information about products made in other countries. In fact, some U.S. data indicate that 10% of people watching TV are surfing on the Internet at the same time (Coffey & Stipp, 1997; Internet users, 1997). Although the infrastructure in Germany allows such behavior, metered local telecommunication costs or cable connection charges prevent German end-users from spending too much time online (see chap. 3, e.g., Table 3.1).

Arrow A symbolizes the *unidirectional* influence of the natural environment on innate individual factors. For instance, a change in climate may alter a society's degree of pigmentation over generations.

Arrow B is *bidirectional*. This demonstrates that approximate factors, such as people's beliefs or attitudes, influence the humanmade environment, and vice versa. For instance, when the steam train was introduced, many people believed it to be evil and resisted its development and use. Hence, the development of the infrastructure necessary to the development of the new technology was obstructed by societal beliefs. Medical technology is another example of the contradiction of societal beliefs and technological advances. It is now possible for a 12-week-old fetus to develop in the body of its brain-dead mother until its life may be maintained in an incubator. Such cases have resulted in discussions between politicians, doctors, churches, and the public about the ethics of such action. The cloning of a sheep in England reported internationally in 1997 led to the scientist being invited to a U.S. Senate Hearing discussing ethical issues and the potential of human cloning. It may be inferred that people's values and beliefs ultimately decide if the medical infrastructure can be used for such procedures. Similarly, the public's attitudes and opinions influence the content (e.g., pornography) and information offered through the Internet.

Arrow B also indicates that economic, legal, and political systems or the humanmade environment can influence people's beliefs and opinions. For example, it is illegal to create or distribute malicious software codes and viruses. In other countries, such acts may not be considered illegal, or are not enforced.

Figure 4.1 illustrates the moderating effects of *cultural derivatives*, as represented by symbols, rituals, myths, norms, and ideational systems. For instance, in the late 1970s, German speaking countries required students who studied business administration and management to demonstrate knowledge about a programming language, primarily Cobol (a norm). In contrast, American universities dropped programming requirements and shifted toward end-user training in applications. Today, German university students may still learn about programming, whereas U.S. business students acquire substantial skills for user applications. This illustrates how changes in the environment (technology and business) can manifest themselves by altering or replacing certain training requirements, whereas new skills may become a norm to secure one's employment after graduation (Gattiker, 1990b, 1990c, 1995). This process may also suggest shifts in attitudes and opinions by educators, employers, and students (i.e., Germany focuses on computing science-type training for management students, whereas the United States gears end-user training toward applications in a LAN and PC environment).

Table 4.1 lists and illustrates the objective and subjective dimensions of culture and in combination with Fig. 4.1, permits us to structure our analysis of cyberspace culture somewhat better, as addressed later.

CULTURE AND TV/CYBERSPACE ENTERTAINMENT

Today's satellite and cable distribution technology has increased the availability of television programming options for households. Particularly, speciality channels (e.g., news, sports, and music) have become quite successful. CNN (all news channel) and MTV (music) were paving the way by being broadcast worldwide. Unfortunately, CNN and MTV offer a fare

TABLE 4.1

Classifying Cultural Variables: Objective and Subjective Culture

Subjective Culture
1) subsistence system (methods of exploitation of the ecology to survive, such as telecommunication infrastructure);
2) cultural system (humanmade environment, religion);
3) social system (patterns of interaction, such as roles and stereotypes);
4) interindividual system (e.g., social behaviors on the Internet);
5) individual system (e.g., perceptions, attitudes and beliefs of what is or is not morally acceptable behavior, such as pirating software).

Objective Culture
1) ecology (e.g., the physical environment, resources, geography, climate, fauna and flora);
2) objective portion of the cultural system or infrastructure (e.g., roads, tools, machines, CIS, and factories).

Note. This list of variables is adapted from Triandis (1977, p. 144) and Gattiker and Willoughby (1993); expansions and additions have been made. In the context of this book, I discuss the objective dimension of culture (e.g., chap. 3 and telecommunication infrastructure) as well as the subjective one (e.g., chap. 4, crime and terror). In this chapter, I focus on the subjective side of culture.

that is clearly influenced by U.S. culture. Accordingly, CNN offers news very much influenced by the current agenda of U.S. foreign policy whereas MTV shows much music based on U.S. teenagers' tastes. Having a fully staffed office in London for its European channel does not remove the U.S. cultural influence. However, across non-English speaking Europe, the dominance of British and American music stars is being eroded. Consumers are increasingly buying local music although pan-European stars are emerging and some are even selling in Latin America.

This apparent paradox has happened during a time whereby geographical boundaries have become less of a hurdle against reaching beyond national markets thanks to technology (e.g., TV and Internet) but, in contrast, cultural differences in tastes such as music have resulted in further segmentation of markets. Accordingly, music channels with a substantial local content (e.g., Via in Germany) and news channels focusing on national or regional issues (e.g., ntv Germany) have blossomed. Moreover, this apparent preference of a substantial portion of the consumer market for local fair in music and TV news is also reflected by cyberspace content. For instance, besides being in German, AOL Germany's content differs substantially from its U.S. counterpart. The German AOL joint venture with Bertelsmann is offering various discussion groups on educational or job-related issues that make up a far greater portion of the total number of such lists on Germany's AOL than is the case in the United States.

This suggests that although culture and cyberspace developments are interrelated, we need to discuss the postmodern condition and its effect on how we perceive and evaluate the world we live in (cf. Sim, 1992). In turn, this might explain why cultural differences have not become less but more distinct in recent decades even though technology may have removed physical barriers (e.g., ease of travel or communication) suggesting an approximation of values in neighboring countries (e.g., EU member states). The modern is an unprecedented effort to explain the world, to conquer uncharted territory and spheres of life while transforming it at the same time. Industrial society is based on technology that is used to improve the allocation of capital and labor (cf. Lyotard, 1984, p. 45).

Table 4.2 defines postmodernism, postindustrial culture, and cyberspace culture. Of course, its appropriateness can be debated. Some would argue that we cannot define cyberspace culture and structure until

TABLE 4.2
Definition of Postmodernism, Postindustrial, and Cyberspace Culture

Postmodernism is a broad cultural condition that characterizes the industrialized world toward the latter part of the 20th century. It entails a crisis in philosophy, arts, and their representation influencing language, knowledge, understanding, and innovation.

Postindustrial culture is shaped by innovation and technology that are applied through our choices made for various technical systems utilized at work and elsewhere. Technical systems are based on research that is no longer driven by proof to win over others but, instead, by performativity (cf. Lyotard, 1984, p. 46; i.e., the best possible input or output relation for its structural features, namely, information and knowledge needed by potential users of the innovation and technology). Hence, technical systems depict the choices made about technology and innovation reflecting our values, attitudes, and beliefs in postindustrial culture overshadowed by the performativity theorem (Gattiker, 1992, p. 304–311).

Cyberspace culture is part of postmodernism and postindustrial culture and, as such, the words, images, communication, and art transcending through cyberspace emerge without rules or categories (see also chap. 7). Instead, end-users search and only after an event has "occurred" a few times, may people be able to formulate the rules of what has happened. Accordingly, cyberspace and virtual reality may not be as clearly definable as suggested in Tables 1.3 and 1.4 as technophiles would like us to believe, and instead, experiences, dimensions, attributes, and rules, as well as moral understanding and justice, are understood according to the paradox of the future (post) anterior (modo).

we have experienced it as a stage in our society's development. Nevertheless, an imperfect attempt, as demonstrated in Table 4.2, may be more helpful than suggesting that communication breaks down and the desire to transcend value judgement is questionable. Lyotard's defiant claim, as follows, does little to help us find common ground in order to facilitate a peaceful coexistence in a postmodern world where cyberspace is becoming an ever more important sphere of our lives.

> When your're trying to think something in philosophy, you don't care less about the addressee, you don't give a damn. Someone comes along and says, "I don't understand a word of what you say, of what you write": and I reply, "I don't give a damn. That's not the problem, I don't feel responsible towards you. You're not my judge in this matter." (Lyotard, 1988, pp. 104–105)

The crisis in philosophy and other social sciences is influencing postmodern culture as represented by technology and art. As such, entertainment has shifted to new genres and forms. Ultimately, entertainment in cyberspace will be a reflection of the facets of television offered on a number of channels. Moreover, as much as television has been influenced by culture as far as content and advertising are concerned, the Internet may also be influenced. For instance, advertising has become an important consideration to pay for the design and maintenance of Web pages. A firm may budget resources as part of its marketing funds for Web and Internet activities whereas providers (e.g., AOL) may try to keep prices for services as low as possible with the help of advertising. Finally, because of the push by television into the cyberspace world, entertainment between TV and cyberspace may become complementary in some markets as the following section suggests.

Talk Shows

Talk shows are representative of postmodernism in today's worldwide audience reached by the media. There are different genres of talk shows ranging from news magazines (i.e., *Dateline NBC*) to sports commentaries (i.e., *Don Cherry's Grapevine*, Canada) to political talk shows (i.e., *Sunday Edition, Canada*). Other hybrid "infotainment" are exemplified in programs such as *America's Most Wanted*, which mixes news with dramatic recreations.

Talk shows (i.e., *The Oprah Winfrey Show*) often involve audience participation. Munson (1993, p. 9) stated that the host's and program's coherence, depth, and logic are secondary to the emotional investment in multiple meanings and lifestyles. He described this as postmodern tendency. Similar to postmodernism, these talk shows substitute "both–and" for modernity's "either–or," and rarely contain the issues and facts necessary to become well-informed on the issues discussed (cf. Munson, 1993, p. 10).

"Like cyberspace, the talk show's audience participation brings about the deeper interpenetration of public and private spheres" (Munson, 1993, p. 151). The new media consumer experiences whole new interactive places that are no less distinctly situational (Meyrowitz, 1985, pp. 124–125) such as reporting from the battlefield or the floor of the New York stock exchange during a financial market crisis. In some instances, news networks' coverage of international conflicts may have crossed the line between entertainment and news broadcasting. For example, during the Gulf War, viewers were able to watch live reports that were often interrupted by the sniper attacks that forced the reporters to run for cover. Reporters were risking their lives to generate more publicity and ratings for the network.

Reality-based television programs have also been somewhat corrupted in the race for ratings. *America's Funniest Home Videos* is an example of viewer exploitation. Viewers are led to believe that the entertaining video clips, which are mailed in by other viewers, represent chance events. Viewers then judge these clips; the winners are awarded a substantial cash prize. According to Munson (1993, pp. 149–150), many videos have been staged and often pain or embarrassment of the person being filmed may be perceived as somewhat sadistic. Such programs encourage audience participation and interaction, which counterbalances the market and its "rationality." In turn, participants feel that their feelings, values, and opinions count.

Cyberspace and Talk Shows

Talk shows and TV are beginning to provide viewers with a link to cyberspace. For starters, most TV shows have an e-mail address or a web page. Some permit viewers to send in their questions for interview participants or guests via e-mail or through chat lines. Since 1995, popular musicians have been using the Internet to reach their fans by participating in moderated discussion groups and answering questions posed by their fans. Popular musicians are also entering cyberspace to promote their albums or tours. Deutsche Welle (German Wave—shortwave radio) claimed in 1996 that its web page had 40,000 visitors per day from around the world all taking advantage of additional programming information and/or materials being offered for reading or downloading (Gerster, 1996, p. 4). Although talk shows and the Internet may overlap to improve their reach to a larger audience, and thus gain ratings, it appears that *cybershows* that can mesh or blur the TV, radio, newspaper, and cyberspace fields (see Table 4.3) are becoming feasible.

These days, even the weather report on some TV channels (e.g., Germany, ARD) offers people the opportunity to visit a web page, send in e-mail, and participate in gametype activities while the weather reporter answers and discusses questions about weather issues with viewers online after the nightly newscast. Of course, the reporter is not just a reporter but a meteorologist by trade, and so, together with other professionals, is able to discuss such issues about weather patterns in depth. But again cultural differences may occur. For instance, research in the United States indicates that the use of a show's web page increases substantially if prominently featured sometime during the show (McDonald, 1997). However, due to me-

TABLE 4.3

Defining Cybershows

Cybershows are an effort by media and viewers or participants to intertwine radio and television with the Internet in order to:

1) reach a larger audience:
   a) possibly beyond the geographical area normally covered by the program (i.e., people who cannot listen to the program on radio or watch it on TV);
   b) by stirring groups that may search for infotainment on cyberspace but not on the radio dial or by surfing the TV channels;

2) encourage the viewer to spend more time with one's favorite show by, for example, visiting the soap opera's Web page for personal news about actors in the show and upcoming stories about one's favorite characters in between shows;

3) defend viewership against another medium (e.g., Web) that could reduce the shows ratings (see footnote 2 in chap. 7).

The structure of this type of genre is still progressing and evolving. With the increased possibilities for transferring voice and visual images through the Internet (see ISDN being 700 times faster than a 14.4 byte modem and cable TV access providing an 80 times faster service than ISDN), cybershows will develop rapidly in offering entertainment and infotainment to new groups of cyberspace, TV, and radio viewers or users. In fact, several soap operas made for the Web only have been on the Net since 1995 and their viewership is increasing.

In the above example, the merging of a TV show with a Web site offers the TV viewer many of the features required to sustain frequent return visits to the show's Web site and, most importantly, to becoming a regular viewer of the show (see chap. 6, also Tables). With the Web site, the TV viewer has the opportunity to stay engaged with the show far beyond its air time and thus become attached to, if not identify with, the show. A scary thought.

tered costs for connecting to and surfing on the Internet, European viewers are rarely on the Internet while concurrently watching TV. Research indicates that this applies to about 10% of TV viewers at any one time in the United States (Coffey & Stipp, 1997). Again, regulation or deregulation as well as costs (objective dimension of culture according to Fig. 4.1), as discussed in part 1 of the book, affect people's behavior as far as cyberspace is concerned. If connecting to the Internet represents a fixed monthly cost and the PC is located in the TV room in one's house, people might simultaneously explore a show's Web site (e.g., during advertising blocks on TV) while following the sitcom on TV. But in other countries, houses may not have a TV room, with the PC being located in one's home office and/or bedroom. Hence, the percentage may look different again for Japan where most people live in small apartments.

As Table 4.3 suggests, the more the Internet is used as an infotainment tool, the more television and other media (e.g., newspapers online) participate. For instance, Politiken, a Danish daily, offers a news summary (in Danish) to interested parties via e-mail for free. The information is detailed enough to help one keeping abreast with news if on the road, while enticing Danish readers to have a look at the full article in today's paper edition. Moreover, cybershows will further blur the private-to-public continuum by permitting viewers to participate more actively (i.e., discuss news or further develop stories) through various moderated or privately administered viewer groups on the Internet.

To further illustrate this, on April 18, 1997 a CBS special, *Cold Case* (www.coldcase.com), used the Internet extensively for its viewers to help with unsolved crimes in the United States. It offered viewers the option to download case files including crime scene photos, incident as well as autopsy reports, witness statements, and detectives' notes as well as the opportunity to discuss the case using the show's chat room. Media hype about the show started right after its official debut during the Spring Internet World convention, March 14, 1997 in Los Angeles.

The idea behind *Cold Case* is to have viewers help police forces across the United States solve crimes. For TV, the show offered an experiment in intertwining the Internet and television. Most important for CBS and its affiliates was that they could claim having done something for the public good by helping solve crimes with *Cold Case* viewers. The show represents continuous innovation by using a new medium combined with a well-established one to increase ratings (especially share for prime time), thereby helping CBS garner more of the advertising dollar. It is also interesting that the Web site was primarily paid for by advertising.

Market Maturation.   Talk shows and cyberspace content or entertainment continue to evolve and regenerate in different ways. Traditionally, audience participation consisted of either a phone-in segment or allowing audience members to ask questions. More recently, the ability to send questions or comments by e-mail or to use one's multimedia workstation to provide the talk show host and audience with one's visual image has mushroomed. Accordingly, newer technologies are engrossed with "older" ones (e.g., TV and telephone), thereby mediating communication and participation between various groups.

This shows that TV has to some degree "invaded" cyberspace by offering its fare on various web sites and by trying to leverage its reach by using the Internet to better link its viewers with shows, news, and other fare offered as part of programming. Hence, most TV stations have their Web sites and advertise them regularly in the hope that viewers will use them as Internet jump-off points (e.g., http://www.nbc.europe.com). In turn, a certain maturity of the market has resulted in brand names coming again to the forefront (e.g., Disney). This is to say that in the past, newcomers may have been charting this new territory with flourish and some success (e.g., Amazon books). Unfortunately for these innovative start-up firms, large organizations with financial muscle, marketing savvy and brand recognition are starting to make their presence felt on the Internet. Hence well-known brands (e.g., Disney) or sheer financial muscle and marketing power (e.g., Motorola) as well as brand recognition in their respective industries (e.g., engineering and Danfoss) help established firms to create Web sites that will attract substantial user traffic. In turn, although some firms such as Amazon books may have created whole new industries (selling books via the Internet), ultimately it may be the established firms who reap a large part of the benefits. One question remains, however; has cyberspace further developed as a cultural genre[1] or is it still a stepchild of TV, techies, or maybe both?

CYBERSPACE AS A CULTURAL GENRE

CIS and the Internet have the potential to change organizational communications' meanings and relations, to support new ways of conducting work, and to foster new genres of media use (Rice & Steinfield, 1994; Sproull & Kiesler,

---

[1]We can conceptualize genres and artifacts as a particular sense of "structure"; the structuring of communication media, such as CIS, TV, or the business letter, through use and interpretation (Rice & Gattiker, 2000).

1991; Yates & Orlikowski, 1992). In turn, CIS, like traditional media, may be altered through institutional and individual transformations (Carey, 1990). Most important is that cultural values, attitudes, language, semantics, moral development, and the interpretation of events as they occur on the Internet can influence the direction we are headed.

Cultural Values

The development of an individual's values is grounded in the culture in which he or she is raised. *Culture*, which is often understood as an analytical variable at a national level, refers to the values of individuals or a group. Cultural values are usually internalized to varying degrees by different members of society. In addition to the dominance of English in international trade (see following section), the convergence theory suggests that with increased development in industrialized countries, people's attitudes may become similar. Accordingly, one might assume that cyberspace consumers or users may display similar attitudes and educational levels. However, research has shown that computer users in two countries that share similar levels of economic development and language (i.e., Canada and the United States) can have different concerns about how the technology invades their workspace, privacy, and quality of work life (Gattiker & Nelligan, 1988).

Research also indicates a lack of agreement regarding the definition of work when cross-national comparisons are made (England & Harpz, 1990). Elizur, Borg, Hunt, and Beck's (1991) data raise some methodological issues regarding the interpretation of England's (1987) research about the meaning of work. If cognitive work values are ordered differently between Germany and the United States, then how can we interpret the finding that Germany has the lowest levels of positively experiencing work and that the United States is in the middle (England & Harpz, 1990)? The ego ideals of Germans reflect a relative emphasis on being critical, informed, and logical. In contrast, the American ideals are to be empathic, perceptive, and genuine (e.g., Gielen, 1982). Such ego ideals or values may affect how Germans respond to a survey, which may possibly account for some of the differences reported by England and colleagues.

These findings suggest that differences in values, beliefs, and opinions may not be easy to interpret and that their effects on technological issues are still unknown. For instance, if Germans value being critical, informed, and logical, then they might be less inclined to be positive toward cyberspace, unless data is provided for the logical assessment of the advantages and disadvantages. In contrast, the American public may be less interested in such data and may focus on the assurances of politicians and experts about safety and appropriateness. This points out that different values and beliefs in separate countries can lead to different assessments and decisions based on one's ethical framework. Accordingly, international agreements about the Internet depend on users having similar values, beliefs, and opinions. If these opinions are different, which is a likely prospect, then we must find a compromise that results in agreement that everybody accepts, understands, and is willing to adhere to (cf. Table 5.7). Reaching an agreement is not a small feat, considering the problems we have protecting the environment and animals from extinction.

Educational Use of CIS.   Educational institutions play an important role in the rapid diffusion of technology for commercial vendors. The number of personal computers available (per 100 people) represents an important statistic. Similarly, when Statistic Denmark announced that by the end of 1998, over 500,000 households of 2,200,000 were connected to the Internet, politicians and media felt that this was an important development. However, as pointed out in chapter 2, over the long term, the use and availability of Internet access to today's youth may be a more accurate indicator of the future use of cyberspace then looking at household access or organizational use.

In December 1995, the software company Oracle announced it would provide free software and Internet access to all 25,000 British primary and secondary schools. Some stated that this would have a profound impact on Great Britain's cyberspace culture and scene (Taylor, 1995); not only does it guarantee the extensive use of the Oracle Power-Browser software but, more importantly, it increases the use of CIS and the Internet by today's students who are tomorrow's workforce. Unfortunately, many schools had not yet taken advantage of this offer in 1998 due to lack of funds for hardware.

Oracle's strategic move, in combination with these research findings, suggests that the Internet will be significantly affected by the values and attitudes of British schoolchildren. Their cognitive and attitudinal outlook (or antecedents) will influence their indirect behavioral tendencies of skirting the law (e.g., software piracy, playing with viruses, encryption and decryption devices, and hacking) and use of the Internet (e.g., time spent, money paid to access provider and telecommunication supplier as well as products and services purchased on average each month). Factors influencing these choices will be the accessibility of the Internet and the possible legal consequences and sanctions children may experience when surfing. In conclusion, the cultural values of tomorrow's Internet are shaped by today's attitudes and the large percentage of pupils surfing the Net, both of which will have a profound impact on how cyberspace culture defines what is right or wrong.

Language   English is the most widely used official language in the world followed by Chinese, Hindi, Spanish, Russian, and French. It makes strategic sense to first write software in English to access the largest market, and then to translate the English version into other languages where markets have the resources and demand for the product. However, different cultural values and opinions about computer games may limit the market for certain games that are highly successful in the United States. Time Warner Germany decided not to translate Hardball, a computer game for basketball. The costs incurred for translation and cultural adaptation would have exceeded the potential sales in Germany.

Although some convergence in attitudes is possible, customers in cyberspace are from very different backgrounds. Because of this diversity, Internet users need to be aware, tolerant, and understanding of the differences and similarities that exist between cultures, political ideologies, and legislation of various countries (Gattiker & Willoughby, 1993). Language plays an important role in the culture of a country; we know that words translated in different languages may not convey the same meaning. In order to reach many consumers around the world, cultural differences must be carefully considered, even if the same language is spoken. Furthermore, in order to reach a large number of people around the world with one's message, the four criteria outlined in Table 4.4 (see p. 94) must be considered.

Worldwide Services.   In 1995, Video Online marked the beginning of a global Internet service provider making most of the information and databases available to consumers in languages including English, French, German, and Russian. To save money, translations were done in Albania. Although they were not perfect, they were and still are culturally sensitive enough to enable the service to reach a far wider audience than it could have by using only the home country's language. Even AltaVista's translator was unable to provide translation for parts of the Starr Report on Clinton, allowing one to make sense of the text translated from English into German, for instance. Nevertheless, the Starr Report made Internet history by being spread like wildfire around the world providing instant details to millions. U.S. providers that are monolingual, and where a translation of a brochure does not yet make one culturally sensitive to anything beyond the continental United States (e.g., Compuserve translating U.S. material for Europe).

TABLE 4.4

What Does it Take for Successful Communication on the Internet?

To assure that one's message can be decoded with ease, and that the message is interpreted accurately by people around the world, some basic factors must be taken into careful consideration. They are as follows:

(1)   Aside from using one's local language, all information on the Internet should also be available in English;

(2)   To avoid misunderstandings between geographical regions due to cultural, religious, or other differences, communication must be carefully crafted. Colloquial writing is inappropriate and humor and jokes have little, if any, place because they may be humorous in one country, but negatively perceived in another;

(3)   One must take into consideration that English is not the first language for many recipients. Therefore, careful crafting of one's communication (e.g., words and expressions used) is necessary to enable people with limited knowledge of the language to understand the message; and finally,

(4)   To assure speed, keep graphics and colors at a minimum, thereby permitting individuals with slow speed modems and limited bandwidth connections from their Internet gateway to read the information with ease (cf. Table 6.4).

Note. These suggestions are often overlooked or simply ignored when using the Internet to communicate with others around the world. For instance, material in Germany or Switzerland is not presented in English (e.g., high speed computer center in Nanno, Ticino has information in Italian only). In other instances, nice pictures are presented (e.g., City of Zurich has a picture of the Limmatquai—downtown area of Zurich available online) that makes the downloading of the information slow and cumbersome from nearly everywhere around the world.

Organizations that provide information on the Internet requiring a combination of audio and video limit the accessibility of that file to only those having multimedia technology, that is, only about 30% of users worldwide in 1998. Moreover, the majority of people having access to multimedia workstations live in industrialized countries, thereby reducing the web page's reach in developing markets.

Pertinent additional information about these issues is provided in chapter 6 (see also Table 6.4).

As Fig. 1.2 and Table 1.1 suggested, the level of reach and the structure of the Internet by large service providers from the United States is high. Moreover, the types of cultures and codes of conduct are grounded in U.S. values and norms (cf. Table 1.1); accordingly, cultural taboos may be violated in other countries by approaching the Internet venture looking at people through U.S. culture-tinted glasses.

Verbal and written communication may be affected by language and the medium being used to communicate. The *flow of experience* (i.e., control of interaction, pace of communication, focus of attention on text and pictures, possibly aroused curiosity and intrinsic interest) has been positively related to perceived interaction with the CIS technology compared to voice mail (e.g., Trevino & Webster, 1992). For instance, in a French company, e-mail messages may be in French. However, for internal communication at its German subsidiary, e-mail messages may be in German. Finally, when communicating with its German subsidiary, French headquarters may use English. Consequently, when communicating with each other, the German and French individuals may be communicating in a foreign language. How this may affect the content, flow of experience, and evaluation of the process by all parties has not been researched; nevertheless, the increased use of the Internet makes this an important issue. Unfortunately, this author is not aware of research that has addressed how language may affect perceived flow of experience of CIS by an individual or how individuals differ in their evaluations based on language and semantics.

Cultural equivalence in research (e.g., Hulin, Drasgow, & Komocar, 1982) indicates that bilingual Puerto Ricans answer the same set of questions differently when asked in Spanish or English. Similarly, this lack of cultural equivalence could affect the semantics and content of communication in different languages. Therefore, if a German or Chinese each communicates with a fellow in their native tongue, then different rituals, norms, and content of communication will be used than if a German and Chinese communicate in English with each other. This further reinforces research findings that suggest that people use different cultural mindsets and frames of reference depending on the language in which they communicate (cf. Hulin, Drasgwo, & Komocar, 1982).

"MTV" Culture and Virtual Communities. If cyberspace develops further as a cultural genre, it will probably be far from being a unified genre. To illustrate, even teenagers chatting with each other on the Internet in Canada develop their own type of language, expressions and abbreviations to communicate. In turn, this makes it hard for their parents to understand the content of any messages they might come across. However, not only parents may have difficulty but also teenagers' peers from other countries whose first language is not English. When my teenage daughter sends some of these chat transcripts to her European friends, they often come back totally puzzled about some of the dialogue or expressions being used by the teenagers across the Atlantic. Being fluent in English does not necessarily mean

one can decipher a message and understand its content without some assistance for decoding the genre and syntax, abbreviations, slang, and much more being used to get one's message across.

Although there will not be a unified youth culture on the Internet as our discussions would suggest, nonetheless, youngsters will be very influential in how the Internet and electronic commerce will develop further. Access at educational institutions and the increasing use of cyberspace to deliver education (e.g., virtual library and university) will make the younger generation Internet-literate and more receptive to its use in various parts of their life (e.g., work and nonwork). But cyberspace may become quite a fragmented place permitting various groups of people to pursue their interests while ignoring much of what is happening around them.

Virtual communities are likely to become communities uniting people from various walks of life and countries sharing an interest. For instance, an individual could be a member of several virtual communities such as one uniting bridge or chess players (entertainment), electrical engineers working in the small household appliance industry (professionals), and tourists traveling to Greek islands (consumers). In each of these virtual communities, topics discussed, information provided, and product or services offered for a fee or for free are of use to members. They all have one thing in common, namely sharing an interest or even a passion on a topic, game, or subject with other members of the virtual community.

Accordingly, virtual communities will become a place whereby people find others who share their interest and may be able to provide help to find one's way in the maze of information and entertainment provided through computer networks. Here, then, cultural differences may be of lesser importance because being a member in a virtual community is based on one important shared value. For instance, this might be becoming the best chess player or engineer for coffee machines as one can possibly be, while keeping abreast of new developments such as regulations and game strategies.

What we don't know at this stage is how such virtual communities will influence social life. Chatting on the computer with one's high school chums in the evening may not affect one's social interaction much because one will see these friends again the next day when attending school. But how about a virtual community in which one is interacting with individuals that one might rarely, if ever, see face-to-face because they might live far away. Moreover, being a member of virtual communities while having unmetered Internet access may reduce one's interpersonal contact with people in one's neighborhood after work because one may "interact" online but rarely, if ever, leave one's house to go to a party or a pub to meet other people. We do know that people who enjoy unmetered local phone calls spend much more time chatting to their friends nearby compared to their peers having metered service. The latter may more likely visit the neighbor instead of chatting on the phone for an hour. On a more personal level, I discovered that while working in Germany and Denmark, I am much more likely to get a private phone call during working hours than when working in Canada and, instead of having a long phone chat, people try keeping phone calls short. Moreover, in Europe, my friends prefer to drop by my house for a chat or cup of coffee rather than using the phone for a one hour long chat as they did in Canada.

How the Internet will affect our social and work lives as far as interacting with other people and building friendships is concerned is still uncharted research territory. Only the future will give us a better picture about this new phenomenon of people being lonely in their geographical location by having few social contacts and friends, while having many friends in cyberspace at the same time. We still know little about how using the phone to chat with friends versus a personal visit may affect friendship and a person's well-being; we know even less about how cyberspace relationships may influence our social and communication skills (verbal and oral).

SUMMARY AND CONCLUSION

In conclusion, researchers and users of the Internet dealing with or interested in cyberspace issues must be aware that: (a) Culture helps us in better understanding the why and how of what people do in cyberspace; (b) Similarities and differences between end-users and countries will be important to understand and explain; and (c) Cultural roots may be reflected in different behaviors and policies affecting CIS and the use of the Internet. The use of an interdisciplinary approach such as that outlined in Fig. 4.1 is a small but important step in the right direction.

This chapter also suggests that use of new CIS technology for entertainment will continue and new genres will evolve. Nevertheless, people who have never met before are coming into contact with one another through CIS. This suggests that although CIS may bring us closer, without understanding one another's cultural background (both objective and subjective, micro and macro), misunderstanding and, possibly, conflicts may arise. Consequently, the challenge for all of us is to keep an open mind and learn more about each other's cultures.

Considering cultural differences and similarities on the Internet, marketing professionals will face some challenges. For instance, although certain advertising strategies may be limited to geographical markets and their printed media, such geographical borders may have little meaning in cyberspace. Hence, a Hindu or Buddhist monk may view an Internet commercial made in Italy differently in comparison to a European viewer. Here, misinterpretations or misunderstandings may occur. And although a company like Philips is able to state in its terms and conditions that Dutch laws apply in case of disputes or liability concerns about its Web site contents and services or products, cultural misunderstandings may still result in miffed web page visitors from faraway places.

5 Ethics and Morals

OVERVIEW

This section gives the reader a framework for studying changes and developments on the Internet with the help and understanding of ethics, morality, justice, and the rights of individuals and groups. Choices made about technology and its use on the Internet reflect our values and beliefs of what is right or wrong and what is morally acceptable. Cyberspace is a medium that transcends boundaries and connects users worldwide. Hence, morality and justice principles from one country may fail to prevail, although U.S. regulators have undertaken attempts in this direction. Success has been limited, and privacy lobby groups, including business, have opposed regulation citing business costs of privacy and rights concerns (Appendices C, D, & E).

Our values, beliefs, interests, and objectives guide which customs, norms, and rituals we adhere to (cf. Fig. 4.1). Based on this cultural framework, we may then perceive certain actions and behaviors as morally acceptable or not (cf. Fig. 5.1). Ethics provide the foundation for our morals, understanding of justice and, ultimately, for the law. In turn, ethics (see Fig. 5.1) affect our cyberspace choices, behaviors, and the resulting technical systems we create (e.g., entertainment with the help of video games or virtual reality; cf. Fig. 1.1 and Table 3.2).

We may still have a long way to go before uniform ethics and morals for the various parties taking advantage of the Internet are understood and, most importantly, abided by. In fact, countries and people may resist such efforts wanting to maintain their cultural heritage and a society's unique-

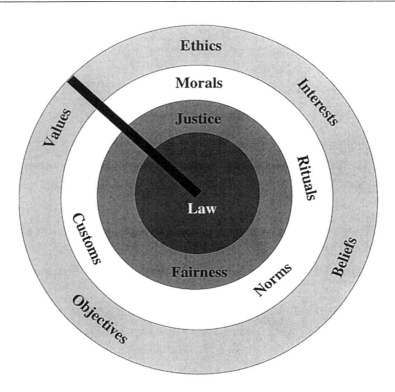

FIG. 5.1.   Human rights and the law.

ness. This will act as a counterforce against the Internet becoming a closely knit community resembling anything near an institution.

ETHICS

The following section discusses ethics in order to provide some conceptual framework and ultimately returns to Internet issues. In today's world, ethics have risen to a new preeminence. They have become an integral part of any medical, law, or business school's curriculum. The interest and attention of many scientists, public policy decision makers, managers, and unions to the field of ethics has spiraled. This has created a plethora of writings, empowering the 1982 launch of the *Journal of Business Ethics*, a journal devoted exclusively to the application of these issues in business. Some individuals state that ethics are assumed to consist of those values permitting people to judge right from wrong (Stead & Stead, 1992, p. 18). Others may simply say that ethics permit people to become more sensitive to the concerns and issues rather than finding the "right" answer.

What are the implications of differences in values, attitudes, and ideals across nations and/or religions for ethics and cyberethics? There is no absolute framework for describing ethics. Religion, culture, and economic development are all factors that separate countries' or regions' values, objectives, and norms (cf. chap. 4). Market competition may be considered necessary for economic growth; hence, plant closures may be accepted as necessary adjustments to changing economic circumstances (i.e., free-trade agreements). Material comforts for a few and the concurrent suffering of many may be acceptable in one country but not in another. For example, the hunting of baby seals may be ethical to a North American Native (e.g., Inuit), whereas an environmentalist in Europe may consider the action to be unethical and cruel. Both have valid viewpoints; nevertheless, a compromise must be found to satisfy both parties.

TABLE 5.1

Defining Ethics and Cyberethics

Ethics are a system of goals, ideals, interests, and values that guide personal behavior in daily life. Therefore, reason given to an action by the individual or observer is grounded in:

(1) understanding the ethical framework adhered to by the individual; and

(2) the individual's use of an ethical framework to deduce the reason(s) causing the subsequent action(s).

Cyberethics refer to the field concerned with the creation, innovation, and application of new technology when cruising on the Internet. This includes decisions regarding the existence, number, and typology of cyberspace-related attitudes, behaviors, technology, and choices culminating in particular systems for work and entertainment. They provide a common theoretical framework for the analysis of ethical aspects of cyberspace technology such as computer-mediated information systems (CIS), telecommunications, virtual reality games (cf. Table 1.4), multimedia and other related areas (cf. Fig. 1.1). Therefore, the term cyberethics identifies general critical concerns regarding cyberspace technology and its application that can be articulated from the standpoint of critical humanist reason.

Note. Cyberethics are merely a subdiscipline of ethics. Hence, ethical and moral principles are applied by a person to manage Internet issues, although familiarity may affect how one decides about a certain behavior being either wrong or right based on one's ethical and moral framework (cf. Gattiker & Kelley, 1999).

The critical reader may reflect on this and conclude that the concept of ethics alone is too abstract to permit the scientific analysis required to make decisions about what is appropriate technology (Willoughby, 1990, chap. 2). Moreover, values and interests between groups may be so divergent that a consensus on what is acceptable may not be possible.

Figure 5.1 outlines the ethical framework for determining the level of appropriateness of cyberspace technology and behavior in a particular context. Ethics (e.g., values, interests, beliefs, and objectives) can be seen as the foundation of social morality (i.e., rituals, customs, and norms) and justice (i.e., what is fair and what is not); accordingly, these are embedded within ethics to help determine the consequences of one's actions and to assess if they are just and fair. As Fig. 5.1 also shows, ethics are based on subjective values that make up our understanding of what we consider ethical; without values, reasoning and justification for behaviors and actions cannot be provided. The following sections discuss the meaning and applicability of ethics with respect to social morality, justice, and fairness. An understanding of why the use of certain technology may be morally acceptable, just, and fair may permit us to determine whether or not we could successfully apply ethical principles to cyberspace, beyond the borders of one country.

Figure 5.1 also illustrates that ethics, morals, justice, and the law are embedded within the principles of rights to which we adhere. In other words, the a priori rights we assume an individual or animal has should be protected with the help of the law. In turn, certain rights may be curtailed to protect other rights. For instance, the current trend in North America toward banning smoking from most public buildings reflects society's ethical and moral understanding that nonsmokers should not have to suffer or experience negative health effects due to secondhand smoke. In turn, one can argue that a person's freedom and ability to derive pleasure from the activity has been curtailed. Moreover, one may not have given any expression of consent to not being permitted to smoke at work. Table 5.5 illustrates that all four principles of rights have been violated. However, because this has been done to protect the rights of the majority and to improve the overall health of the population, limiting the rights of smokers appears justified. This clearly illustrates that various rights must be balanced and weighed against each other; immediate versus long-term benefits must also be assessed.

Internet and Culture.   One of the concerns governments may have is that the Internet cannot be as easily regulated as may be possible with broadcasting, whereby one requires a license to broadcast on a particular frequency. Running a radio or TV station requires the firm to meet local guidelines, such as broadcasting a certain percentage of local movies during a week and making public announcements. Distributing an electronic newsletter or offering a chat line on the Internet does not require the "broadcaster" to have a license. The difficulty begins when the newsletter is distributed around the world or when chat participants come from various countries. The local censor may no longer be able to stop certain information from reaching minors or, in the case of politics, local voters. What may be considered indecent in the United States may raise little, if any, eyebrows in Scandinavia.

How we define privacy is part of our culture, this is to say what we value and feel is a person's right (see also Appendix A for a definition of privacy). How much and what kind of our personal information should become part of a database may also be rooted in how we value our privacy and thus what we consider ethically acceptable and what not. For instance, should my pizza-ordering habits be available on a marketing database purchased by a local firm? However, if I do shop on the Web site of a firm located in a country where limited privacy laws are enforced, my shopping habits may become known to many firms interested in such data but not located in my home market where such information might be confidential.

This illustrates that although ethics is influenced by our culture (e.g., what we value and what we find unethical), the reach beyond geographical boundaries may force us to deal with different interpretations of ethics when using the Internet. Not only will different languages make it difficult to find a consensus because cultural nuances and upbringing will result in people behaving differently but, as illustrated earlier, even teenagers in different countries may differ in how they interpret messages.

SOCIAL MORALITY

Ethics are defined as the "higher order" that belongs to every culture or nationality (Conger, Loch, & Helft, 1995) and represent the goal(s) and ideal(s) of every culture while depicting rule(s) that guide(s) personal behavior in daily life (Gattiker & Kelley, 1995). On the other hand, morality provides "an impartial constraint on the pursuit of individual interests" (Gattiker & Kelley, 1999, p. 24). Miller, Bersoff, and Harwood (1990) regarded morality concerns as "based on objective obligations, independent of social consensus or personal preference and as legitimately subject to social control, rather than as the agent's own business." Turiel, Killen, and Hellwig (1987) pointed out that in all cultures, moral issues involve specific topics like justice, rights, or harm, but the cultural-specific rules differ.

The word moral is derived from *mores*, a Latin term meaning customs. As such, morals are grounded in ethics. Moral capacities are subsequently developed toward those individuals and things in which one takes a special interest (e.g., Goldman, 1980, p. 40). Consequently, moral behavior develops outward from intimate relations within the family to one's friends and then to strangers within acceptable social limits (see also Fig. 5.1). Goldman (1980, p. 5) suggested that without human intimacy or an interest in other individuals, one may be unable to develop moral behavior at all.

Gauthier (1986, p. 59) maintained that although an individual's values are subjective and relative, they are proper because they are preferred. As such, a computer with Internet access provides its owner with the freedom to produce and receive data and information relatively conveniently regardless of the geographical location of either the supplier or the recipient of data. Cyberspace users have a preference for such a choice. The question here is: how do rationale principles held by a computer owner serve as constraints when making choices among possible alternatives? These particular rationale principles are our moral principles.

Moral principles guide our public and private conduct. Difficulty arises when an individual does not appear to have an interest in other individuals' limits or a misunderstanding occurs regarding acceptable limits. Much of what happens on the Internet is not limited according to cultural norms and understanding (e.g., Gattiker & Willoughby, 1993). Hence, cultural norms may still be evolving whereas various signals (e.g., nonverbal communication) may not provide the individual with the guidance one needs to fit in. For example, certain norms have been developed regarding phone use (i.e., making a proper introduction of oneself when calling another party). In contrast, when communicating via the Internet, it is not as easy to determine which norms should be followed, because they are not developed. Communication through the Internet, compared to face-to-face social interaction, is more constrained, unless video-supported communication is used. Hence, social complexity is reduced compared to face-to-face communication. Furthermore, this can result in interaction difficulties (e.g., misinterpretation of communication or reading between the lines; cf. Fig. 1.1).

An individual is a member of the cyberspace community and is circumscribed by neither geographical nor time boundaries or constraints (e.g., around the world, 24 hours each day). Accordingly, people from various cultures, countries, and communities may now interact with the help of the Internet.

Table 5.2 provides a general definition of social morality. Professionals may adhere to the moral standards set by their employer, by their profes-

TABLE 5.2

Defining Social Morality and Related Terms

Social morality is concerned with those who choose, execute, and are implicated in the consequences of their deeds and choices. Morality is traditionally understood to involve an impartial constraint on the pursuit of individual interests.

Moral constraints are generally accepted because they foster interaction between parties within an orderly and stable framework.

Agreed mutual constraint is the rational response to certain structures of interaction between individuals pursuing their own interests.

Note. In the context of this book, we primarily focus on how we can pursue cyberethics while adhering to the generally accepted moral constraints of society.

sional code of ethics, and by society. Each profession's moral principles guide the conduct of its members. Morality arises from the application of the process of exchange. For instance, the individual benefit of neglecting to report all income to the tax authorities puts individual interests at odds with those of society. The mutual benefit of an individual being fully taxed requires an agreement between both parties. Although the government receives a share of one's earnings via taxes, it provides the individual taxpayer with a social infrastructure.

Ordinary morality deems our behavior morally acceptable as long as we meet its demands. There is often conflict between one's own interest and the promotion of the common good. The challenge of morality requires one to do as much good as possible. For instance, when one purchases a computer software game, he or she may feel that the money could have been spent toward the betterment of society. However, unless we know about, and understand the existence of, a moral requirement to promote the overall well-being of society, and are aware of a duty to be attentive to the needs of others, it is easier to live according to the pursuit of personal good. The moral dilemma between purchasing a computer game and donating the money to a charity does not affect most of us, because the pursuit of personal interests is paramount.

Internet and Culture.   There might be a conflict between a person's own interest and the promotion of the common good. This could result in the individual purchasing a software online from a vendor abroad. The individual would save costs in comparison if one were to purchase this software in a store (e.g., due to lower overhead for firms doing business with an electronic or virtual store front only) but also saves value-added tax (VAT) or sales tax charged if the software is purchased locally. Many countries' tax codes specify that even services or products purchased electronically from abroad are subject to local excise and sales tax(es). However, people may not report such purchases but instead, may pocket the savings. Individual interests overriding society's has always resulted in some services not being reported to tax authorities. For instance, a person may do gardening or housework without reporting income from such activities. In turn, the individual may offer customers a cash and a check price, thereby permitting the customer to participate in the tax savings. Although this might be perceived unethical, nevertheless, some people will continue to avoid taxation wherever they can by paying cash for some services and/or not paying taxes on products purchased electronically abroad.

Although software piracy (i.e., percentage of total software used illegally) is considered a serious problem by the industry, it must not be of great concern to end-users who have been continuing using pirated software for various purposes. Even if we disagree about how much of a problem software piracy is, it appears customary to have a few software packages on one's home computer for "testing" purposes, naturally. Particularly, games are passed on between users freely and without thinking about possible copyright violations by doing so. Here, individual interests override the software industry's or society's interests. The issue is whether an individual's pursuit of his or her interests to obtain software as cheaply as possible should or should not be curtailed in order to protect others' copyright and property rights. Most local laws would suggest the latter, but software piracy is often perceived as cavalier delict. Other examples are individuals who have no difficulty in signing up with an Internet service provider such as AOL to take advantage of a bonus offer (e.g., 50 free hours and/or 3 months unlimited service for free). As soon as the bonus has run out or has been used up, however, the individual will sign up as another person taking advantage of the next offer. Is such behavior morally just because one person's free ride will have to be paid by the rest of us? This illustrates that we have to also address the social justice issue to determine how much an individual's interests may violate society's or vice versa.

SOCIAL JUSTICE

Internet users may value their freedom to surf in cyberspace for divergent interests; that is, to send and obtain data or information and entertain themselves. From a moral standpoint, an extremist could argue that one may have to avoid cruising on the Internet in order to decrease the number of users during certain times when traffic peaks. Subsequently, a problem arises when a minimalist refuses to accept this moral outlook and continues to make extensive use during these times. Can we enforce a limit on Internet usage without raising concerns of unjust treatment? One group's freedom to cruise on the Net may infringe on another's right to fast responses through the Internet. Accordingly, the development of a just society requires that the rights secured by justice are balanced among the various groups' interests. Freedom, then, is a relative term. One person's freedom to watch pornographic material on the Internet may limit another person's right for a pornofree environment.

Flew (1971, p. 111) pointed out that we can speak about the concept of justice only if we appeal to some "standpoint or principle independent of particular individual or group interests and wishes." To this end, the artificial distinction between us and the others should be avoided. Accordingly, we should look at American and German Internet users not as separate groups, but as one interest group whose divergent values may be affected by imposing a universally enforced limit to reduce overcrowding of the Internet.

A counterargument to this would be to let pricing mechanisms take advantage of the free market instead of imposing a limit (i.e., higher prices are charged during peak hours). This example is comparable to Singapore's policy of charging variable fees to drivers entering various parts of the city according to day and time. The car example would also illustrate that most of us have gotten used to the convenience of driving whenever and wherever we wish to. Similar to today's telephone charges that are lower during off-peak hours, so are rates paid by advertisers during TV prime time between 19:00 to 23:00 hours. Similarly, variable costing for e-mail and Web surfing could be used to reduce bottlenecks. Accordingly, when traffic is assumed to be highest, such as during lunch hours from 11:30–13:30 and after work from 16:00–23:00 hours, when traffic is often high in local markets, higher charges would apply. But this might meet resistance from businesses that enjoy fixed line charges to connect 24 hours a day to the Internet. Various pricing systems would increase costs for business users beyond possible volume charges from the telecommunication or cable connection provider and incurred based on the amount of traffic a Web site enjoys. But already today, some North American Internet providers apply variable pricing in some form or another for their private clients. For instance, users with unmetered or unlimited Internet access may be restricted to about 2 hours of access between 17:00 to 23:00 hours each day, unless they are willing to incur additional charges.

This illustrates that what is just may change over time and may depend on how each individual or social group is affected. If change is perceived as negative, it may trigger resistance. For instance, private users with unmetered Internet access but with restrictions experienced during peak hours or with difficulty to reach the provider may simply switch to another provider if given the opportunity. Any definition of justice must reach beyond individual interests, hence, organizational and individual interests should be balanced. Rawls (1971) saw the concept of justice being defined by "the role of its principles in assigning rights and duties and in defining the appropriate division of social advantages. A conception of justice is an interpretation of this role" (p. 10).

Assigning rights and duties, while dividing the social advantages in an international context, necessitates international sovereignty, which may be difficult for some countries to accept. For instance, if the United States enforces a particular standard for data encoding (e.g., the debate about Clipper and Capstone Chips during the latter part of the 1990s), other countries will be affected. During such discussions, U.S. politicians and experts usually refer to the domestic issues and concerns, whereas international ones are often ignored.

In Rawls' definition, government plays an important role in ensuring that the principles of justice are realized within society. For Nozick (1974), the role of government is limited to providing protection, and the complete principle of distributive justice would simply say "that a distribution is just if everyone is entitled to the holdings they possess under the distribution" (Nozick, 1974, p. 151).

Although it appears that Rawls and Nozick have incongruent views on justice, both of their understandings of the concept are based on a system of values. In both instances, the ethical framework entailing an individual's and society's values, objectives, and interests is of vital importance. On what values do we base the process of appropriately dividing the social advantages—ours or theirs—or can we reach an acceptable compromise? If we resolve this question, the distribution will be just, because everybody is entitled to the possessions they hold. Defining rights and duties (see Rawls) and determining what is just requires that we define, assess, and discuss the ethical framework on which we base our state and society. If ethical frameworks differ between groups, conflicts will occur. Moreover, as Gauthier (1986, chap. 2) pointed out, an individual's or a society's choices are based on considered preferences, and reflect the best estimate of choices that would pass the tests of reexamination and experience. Social morality by agreement provides the constraints of these choices to permit the pursuits

TABLE 5.3

Defining Justice

The **essence of justice** is to appeal to principles logically independent of any particular individual's, group's, or country's specific interests, beliefs, objectives, and values. The two major principles of justice are that it assigns rights and duties and that it defines the appropriate division of social advantages (see Rawls, 1971, p. 10).

The **principles of justice** are founded in:

(1)   the ethical framework of values measuring individual and societal preferences; and

(2)   social morality, which represents voluntarily accepted constraints developed from the rational agreement of equals.

of those preferences based on a person's individual values. However, neither Rawls nor Nozick addressed this more practical issue of discourse. Table 5.3 provides a definition of the term "justice" that encompasses its moral and ethical foundations.

Values measure individual preferences, and are therefore subjective and relative. We use computers to achieve objectives that inherently represent our personal preferences within the moral constraints given by society (see Table 5.1). We can only agree that these objectives give technology a reason for its existence (e.g, Hersch, 1978, p. 139). However, it is the ethical framework and the values it entails that help us to understand and define just and unjust (see Table 5.3). For instance, our value for communication will explain why we may have chosen a certain objective (e.g., completion of a fibre optics network) with the support of computer-based technology.

Because too much information was leaked through the Internet about former president Mitterand's sickness, the French government may feel it is just to regulate the Internet. In contrast, whereas Chinese officials were outraged, the West appreciated the opportunity to receive information via the Internet about China's Tianamen Square massacre in 1994. To achieve justice, a system that permits the flow of information regardless of geographical location and preferences of the current government must be designed. Hence, there are two choices for Internet regulation. Either the Internet and its contents continue to be unregulated or international sovereignty is introduced, which would enforce a worldwide standard.

Internet and Culture.   As the previous discussion suggests, what is fair and what is not is rooted in a country's culture. As such, chapters 2 and 3 discuss what and how access to the Internet is being provided to the public. One of the arguments I put forward is that Freenets should be replaced by SaveNets, which reduce people's dependency on the public coffers for providing a service at subsidized costs or even for free as far as Internet access from home is concerned. Naturally, some may argue that the impoverished in industrialized countries should receive Internet access for free or as part of social welfare. Because poverty is, however, a relative term (Sarlo, 1992), it is questionable whether social justice in a country necessitates Internet access and/or a telephone service for free or at a subsidized rate from one's home if one is considered poor.

Assigning rights and duties and the appropriate division of social advantages as far as information and knowledge access is concerned is quite difficult. However, it seems apparent that the Internet, especially with the help of the Web, provides users with a vast array of options for obtaining information and securing additional knowledge about a vast number of subject areas. Access to information and knowledge sources is the Internet's major contribution. Naturally the Internet also provides the user with an easy, cheap, and rapid way of communicating with others via e-mail for private and business-related needs.

Internet and Skills.   Globalization of business has forced firms to find new ways for lowering their wage costs. This is often accomplished with hiring people part-time or on a nonpermanent basis. Often, however, such workers are not given Internet access. In deregulated markets (see chap. 2) such as the United States, this is of little consequence, because many of these employees will have Internet access from home for a reasonable monthly fee including telecommunication or cable access charges (see chap. 3). But in some countries, where a limited number of private households have access to the Internet from home, exclusion at work may further limit these employees' future employability. Acquiring and regularly updating Internet and CIS skills is becoming a prerequisite for many jobs. Until now, dual labor markets have differentiated between workers with more secure jobs enjoying a wide range of benefits (e.g., pension and profit-sharing schemes) and others in nonpermanent positions (e.g., employment limited to a specific time frame). However, labor markets may now also begin to differentiate between Internet literate and illiterate individuals as illustrated in Fig. 5.2.

The x axis in Fig. 5.2's cube assumes that we have two levels of employment whereas the y axis outlines the level of Internet and technology skills (see Appendix A for a definition of the term). At the top, the employee's self-initiative is taken into consideration. Every employer wants workers in

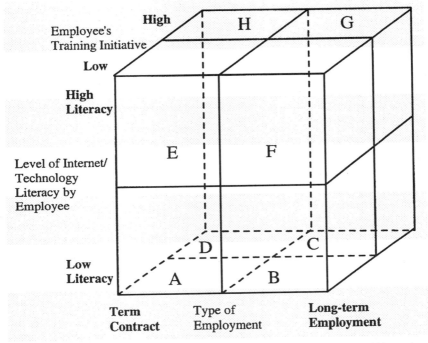

FIG. 5.2.   A model for classifying Internet/technology skill levels. Although type of employment by the employee affect skill levels due in part to organization-sponsored and initiated training, employees may improve their skills through continuous education and through their own efforts by attending courses after work that are not necessarily paid for in full by the employer.

Quadrants G and H but in some instances, employees may show little initiative and may feel that it is the organization's responsibility to provide them with the training needed to keep abreast of new developments (Gattiker, 1994). Unfortunately, although labor and union contracts may assure some investments made by the firm in its workforce's human capital, it is ultimately in the employee's best self-interest to show initiative and keep abreast new developments. In turn, as suggested by human capital theory, the employee's skills may allow him or her to switch employers if necessary (e.g., plant closure) or desirable.

   Although globalization and its pressure on firms to lower costs is nothing new, the Internet has resulted in workers becoming more aware of how the world can be unfair by putting them against each other. Moreover, except for public organizations, most jobs in manufacturing or services can be outsourced to markets abroad and the Internet has made it easier to transfer data and information between plants. Finally, the distribution of new information and knowledge has become easier, faster, and cheaper via the Internet. Hence, the Internet is making it more difficult to keep abreast of new developments without having extensive access to this rapidly developing technology and resulting in changes for one's work and nonwork lives. Most important is that school pupils are able to take advantage of this technology to acquire the skills while becoming familiar and comfortable with the Internet and related technologies. Lack of access during one's education may become a further dividing line between industrialized and developing countries. The latter may be unable, due to the lack of appropriate infrastructure such as electricity, to provide all their school children in rural areas with Internet access (e.g., virtual school library). In turn, students in developing countries may not have as easy an access to know-how and information compared to their peers in industrialized countries, which appears unjust and unfair.

THE RANGE OF RIGHTS

A further difficulty for achieving a just system of information exchange on the Internet is the differences in values (see chap. 4), which, naturally, can lead to diverse behaviors that hamper the potential for progress on cross-national issues. Accordingly, we must play our "cyberspace cards" right to assure justice and fairness for many. Our rights correlate to the active or passive duties of others. If we assume that we have the right to use the Internet in a certain way, this implies a norm that, in turn, imposes upon others the duty of permitting us to do what we want with the technology (build wireless communication networks around the globe; cf. Nino, 1991, pp. 25–29).

Table 5.4 provides some definitions and outlines the requirements for meeting the principle of fairness. The principles of personal autonomy and the hedonist are aggregative principles, insofar as "they assign value to states of affairs and to the goods instrumental to them, regardless of how they are distributed among individuals" (Nino, 1991, p. 186). Accordingly, they establish the content of basic individual rights. The principle of inviolability means that personal autonomy and the hedonist principle cannot be pursued without some moral constraint (i.e., taking away from others while limiting their rights may be unjust).

We assume that the rights described in Table 5.5 are basic rights (e.g., freedom of expression) and a priori. However, conditions of moral discourse, behavior (a priori basic rights), and cyberspace rights presuppose

TABLE 5.4

The Principle of Fairness in a Just System

The **principle of fairness** holds that an individual is compelled to do his or her share as defined by "the rules of an institution" when two conditions are met:

(1) the institution is just (or fair), that is, it satisfies the two principles of justicea; and

(2) one has voluntarily accepted the benefits of the arrangement or taken advantage of the opportunities it offers to further one's interests" (Rawls, 1971, pp. 111–112).

"**Justice** tends to be the salient principle, facilitating acquiescence in third-party interference, when conflicts arise out of prior conventions, agreements, contracts, and attempts at their enforcement" (de Jasay, 1989 , p. 219).

"**Fairness** is salient for social choice, when conflicts cluster around problems of 'who pays what' and 'who gets what' in the absence of prior agreements, calling for resolution of the distributed implications of coerced coordination" (de Jasay, 1989, p. 219).

[a]The two principles are defined in Table 5.3.

TABLE 5.5

The Range of Individual Rights

A person has the right to:

a) freedom to change and realize personal ideals (principle of personal autonomy);

b) some pleasure and the absence of pain (hedonist principle);

c) not be deprived of either a or b for the benefit of another individual (principle of inviolability); and

d) make decisions and expressions of consent (principle of dignity).

Note. The above is derived from Nino (1991, chaps. 4–6). These rights are interrelated with each other. The principles of dignity and autonomy are related because what makes consent and decision morally relevant is that we respect them. We also allow individuals to take external circumstances into consideration (see also The Elements of Rights outlined in Table 5.6).

discourse and democratic procedures to determine what these rights will be. Through continuous discussions, rights may affect or be affected by cyberspace choice and practice as well as by the technical systems prevalent in our lives. Sometimes, this may mean that coercion is justified if it is based on protection of the rights of the people. When there is a chance that such rights of the people are being endangered, it may be ethically and morally justifiable and fair to impose protective measures.

Freedom of Rights.   The challenge for each individual striving to meet his or her desired level of freedom is to define the elements needed to realize one's rights. For instance, enjoying news groups reflecting radical religious or political views offers the Internet user freedom of information. However, one's right to such information may also conflict with another's right to religious freedom. If cooperation occurs by setting guidelines about what content a Web page may have and how we may communicate or express emotions, everyone's freedom is somewhat limited. Individual behaviors lead to observable outcomes. However, a person's perceived optimal degree of freedom as a subjective state can only be known if we can explain his or her motivations for making the choices necessary to realize his or her rights. Conflict arises from the need to assess and measure the freedoms of others and agree on the extent of their rights. The latter are based on the range of individual rights (see Table 2.6), and we must carefully determine how to take advantage of these rights to ensure that the system remains fair and just (see Table 2.5). Table 5.6 outlines the three elements required to assess whether one has the right to pursue a certain action.

TABLE 5.6
The Elements of Rights

An actor has a right to:

(1) choose or reject a behavior (X);
(2) an injunction protecting her or him from either interference from others or lack of support if she or he chooses to pursue or not pursue X; and
(3) an enforcement privilege permitting the agent to enforce the injunction.

Moral constraints may influence an actor's decision about point (1). Hence, the actor may feel that he or she has only a minimal right to pursue X and may permit him or herself to be influenced to change his or her behavior.

Note. Without adherence to moral constraints observed by all parties, it will be difficult to safeguard society against the undesirable effects of CIS and, in particular, the Internet [cf. point (2)]. Table 5.5 defines the range of rights we should adhere to and the four principles on which we base the above elements. If a person rejects a behavior or decides to pursue it, one should make sure that neither of the four principles in Table 5.5 are violated in any way. In turn, a combination of the principles of rights (Table 2.5) and the elements discussed herein, are affecting the cooperative social arrangement needed for securing cyberspace/Internet rights and responsibilities (Table 8.2).

As Table 5.6 suggests, an individual's freedom to pursue a certain action or behavior is constrained by his or her rights and by the rights of others. The following example may be useful in determining why an individual should have a particular right in the first place, and where his or her rights should be subordinate to those of others. If party B has a partial right during a negotiation situation, he or she may influence party A to refrain from exercising the right to pornographic or indecent literature.

Obstacles Facing Cooperative Arrangements.   One of the challenges of forming a comprehensive social contract that will work is that people may agree to something, yet behave in a way that contradicts the agreement. An example is the issue surrounding pirated software. The industry claims that billions of dollars in revenue are lost every year in the United States and Germany alone. Surveys of users indicate that copying pirated software for testing purposes has become acceptable (e.g., Gattiker & Kelley, 1994). If such judgments and behavior become socially accepted, then they start constituting part of positive morality (cf. Nino, 1991, p. 65). Consequently, laws have little meaning unless the public's code of ethics and understanding of morals and fairness result in acceptance and adherence; tolerance, respect, and support of individuals who do so; and enforcement of these laws by each one of us.

If we agree that a certain behavior is important to all of us, then we must enforce a rule and make sure that people in violation pay the appropriate fines (see Table 4.4). Failure to do so will reinforce violation of the rules and we might as well abandon them. The previous example also illustrates that personal interests (e.g., enjoying the benefits of software without necessarily paying the royalty fee) tend to override those interests that promote the good of society.

Internet and Culture.   It may be appropriate to restrict children's access to the Internet with the help of software, which does not permit connecting to certain World Wide Web (WWW) sites from a particular personal computer (offered in the United States to customers since 1995). This restricts the rights of free expression (i.e., the person offering such material on the Web). Accordingly, one individual's personal right is curtailed in the hope of protecting a child. Hence, the distribution of pornographic material usually requires careful assessment of potential violations of legislation in some countries. From a moral standpoint, an extremist could argue that one may have to restrict the use and content of the Internet (e.g., pornography) in order to reduce the chances of children being exposed to pornography. Subsequently, a problem arises when a minimalist refuses to accept this moral outlook and continues to read and distribute pornography through the Internet.

Can we enforce a limit on the number of sites on the Web offering pornographic material, without raising Internet users' fears of being unjustly treated? One group's freedom to watch pornographic pictures may infringe upon another's right to a pornofree environment. Accordingly, the development of a just society requires that the rights secured by justice are balanced among the various groups' interests.

Another concern is how people's rights for keeping their conversations via the Internet private might be violated by an encryption regulation that attempts to exclude people from countries other than the Unites States or Canada from the latest technology (see chap. 2). The principle of fairness as outlined in Table 5.4 may have been violated by having U.S. regulators include and exclude some parties from using the latest technology without ever having been asked about their preference. Similarly, France's requirement that encryption keys need to be deposited with authorized French parties may violate another country's citizen's rights. To illustrate, when communicating with a U.S. friend traveling in France while using noncertified or unapproved encryption technology without necessarily being aware of it, French laws may be violated by both the sender and the recipient. Less democratic governments than the one in France may misuse such harmless violations to stop a person from entering or leaving the country.

Developing a Comprehensive Social Contract

This suggests that the lack of a clear understanding of people's rights and responsibilities across nations in addition to legislation and regulation that lacks harmonization even within trading or political unions (e.g., NAFTA and EU) may further exacerbate the problem. In order to protect most Internet users from having their rights violated, we need a comprehensive social contract for CIS and the Internet. Moreover, protecting Internet rights in one country may not suffice. For instance, privacy laws protecting individual rights to access, correct, or eliminate information from databases in one country may be circumvented by creating, upgrading, and applying the database from abroad. Only international cooperative agreements will secure Internet rights. Table 5.7 provides the three conditions that must be inherent in a contract to make it comprehensive. Principles (cf. Table 5.4) and conditions of the cooperative process must be met in

TABLE 5.7

Form and Content of a Comprehensive Social Contract
for Technology and, in Particular, for Cyberspace

Three conditions must be met to make the social contract comprehensive:
(1) universal acceptance of the rules pertaining to **Internet/cyberspace rights**;
(2) acceptance of the enforcement clause that permits the agency to suppress any behavior contrary to the social choice; and
(3) consent to the enforcement clause.

Note. This pertains especially to content of messages and the ethical and moral issues that extend beyond national borders (cf. Table. 6.1). Making a social contract comprehensive also requires the assumption of an impartial standpoint.

order to discharge one's Internet rights in an ethical way while, most importantly, meeting society's moral and legal standards.

Unfortunately, at this time it does not appear that we are developing a comprehensive social contract for cyberspace and the Internet in some countries. If efforts are undertaken, each country appears to pursue its individual interests driven by domestic political and economic agendas, interest groups, and reelection campaigns. In cases where international agreements are pursued (e.g., privacy protection in EU countries since October 1998), their harmonization with other countries (e.g., between EU and United States) is not being pursued unless trade or other conflicts (e.g., political and national security) are looming on the horizon.

During March 1997, the U.S. Supreme Court also heard arguments involving a new law that wants to ban anything from the Internet that depicts or describes in offensive terms, as measured by contemporary community standards, sexual or excretory activities or organs (it does not target obscenity or child pornography that are already illegal in the United States and many other countries). However, it is simply impossible for the vast majority of Internet users and service providers to distinguish between adults and minors in their audience and, instead, parents should exercise parental responsibility in preventing children from accessing indecent communications, including the ones posted abroad (Vicini, 1997). During the hearing on Wednesday, March 19, 1997 at the U.S. Supreme Court, Justice Stephen Bryer asked whether the government might prosecute a group of high school students discussing their sexual experiences in a chat room on the Internet. If, in June, 1997 the Supreme Court had not rejected the proposed indecency law, this would have made a large number of high school students guilty of a federal crime by discussing sexuality issues on the Internet.

The rejected indecency law in the United States exemplifies how difficult it will be to develop a comprehensive social contract that defines people's rights and responsibilities (e.g., for trading with certain countries and
obtaining "indecent" information on the Internet). It is unlikely that countries will agree on "minor" issues about indecency or extremist information
(e.g., right wing material cannot be posted in Germany, but can be in the
United States). But things might come to a head (e.g., EU vs. United States
on privacy legislation) until one party blinks first and gives in. This international game of "chicken" is played out in various contexts as countries or international organizations use their leverage to get what they want from
other countries. The EU argued against the Burton-Helms law in front of
the World Trade Organization (WTO) in Geneva, whereas the United States
was trying to make its case for the law by pointing out that it represented an
internal security matter and thus did not fall under WTO jurisdiction. Similarly, if the decency versus indecency matter on the Internet does not get resolved satisfactorily beyond the U.S. border, the likelihood that foreigners
will become criminals under U.S. federal law is a real concern (e.g., discussing sexual adventures between a U.S. and a French high school student
in a chat room offered by a French Internet service).

Internet Rights and Responsibilities

The previous discussion illustrates that because our efforts for international
cooperative agreements are dismal in many areas, it does not look like progress will be made by politicians and international bodies unless Internet users and interest groups put some pressure on their elected officials to make
it happen. Being clear about our rights and responsibilities about CIS and
the Internet might be another important step toward a cooperative social
agreement beyond national borders.

De Jasay (1989 , p. 21) suggested that to distinguish between our rights
and freedoms and those of others, we must define the boundary between the
two, which remains blurred. We do not know where German cyberspace users' rights and freedom to receive information from U.S. users about political groups (e.g., Neonazis) are violated. Nor do we know how we can
contain U.S. users' right to such information under the freedom of information act in the U.S. Instead, the minimal explanation for divisive social contracts in which there is cooperation but little acceptance in domains such as
the Internet may be the only feasible alternative we have.

When we discuss general rights in cyberspace or society (e.g., if we are
poor or rich in relationship to receiving welfare vs. paying taxes), we often

appear to ignore the importance of responsibilities we acquire with any right. Hence, the right to privacy or to data protection and safety does not mean we can simply avoid responsibility in what we do with and about data (see Privacy Code—Appendix F). Instead, each individual is responsible for protecting our rights and pursuing them within the moral constraints we adhere to in a society.

Table 5.8 outlines some of the rights and the responsibilities for the Internet user. The rights and responsibilities in Table 5.8 are not listed in any particular order of importance. If we pursue and respect the points outlined in Table 5.8, we, as citizens and Internet users, cannot defer to the government and its representatives to protect our rights. Most importantly, we must carry a large share of this responsibility.

TABLE 5.8

What are my Rights and Responsibilities in Cyberspace,
Virtual Reality, or When Using the Internet?

A comprehensive social contract must meet the three conditions outlined in Table 5.2 to attain for each individual the cyberspace rights and responsibilities to:

(1) create cyberspace products and participate in processes that harms neither oneself nor others;

(2) work and live in a clean (issue of information content), safe (issue of virus writing, damages to data and property) environment;

(3) know and be able to freely discuss the costs and benefits of new Internet development;

(4) participate in a democratic decision-making process concerning Internet development;

(5) evaluate positive and negative effects of possible applications and uses of Internet technology;

(6) be provided with accurate information about cyberspace technology and be willing to carefully study this information to be able to make informed decisions;

(7) be able and willing to gather, analyze, interpret, and discuss any appropriate information about CIS and the Internet so as to make informed decisions; and

(8) refrain from projecting our morals and ethics on others who, because of less wealth, may be unable to resist our attempt to impose our values and morals upon them, thereby very likely violating their rights.

Note. Some of the points outlined here cannot be answered unless we know and understand cyberethics (see Table 5.1) and, in turn, the range of individual rights (cf. Table 5.5) should guide our discussions.

In summary, this section outlines that one of the biggest challenges remaining for all of us is to develop a comprehensive social contract for cyberspace and the Internet, whereby we determine the parameters for the rights and responsibilities each one of us should have in taking advantage of the Internet. In order to continue the betterment of the many and not the few, we must pursue the responsibilities and rights we have in regard to the Internet with caution and vigor. Interest groups, citizens, and users alike must participate in this process and take responsibility for their apathy, behavior, and adherence to the rights and responsibilities pertaining to CIS and, in particular, the Internet.

SUMMARY AND CONCLUSION

Ethics, morals, justice, and the law are embedded in our rights. Without understanding the principles of rights (cf. Table 5.5) and their elements (cf. Table 5.6), it may be hard to follow Internet choices and practices. For instance, in most democratic countries, it is believed that people have inherent rights. These rights are usually reflected in those countries' constitutions (e.g., freedom of speech, right to privacy). In other countries, these rights may not be considered inherent and instead, they are justified as reasonable by explaining certain practices to be the means for serving some interest or benefit. However, we assume that basic rights are inherent and a priori. If these rights are constrained, public protest against nuclear energy or any technology may not always be tolerated. Even democratic systems (e.g., Germany and Switzerland) have had problems tolerating dissonance against certain technology choices, such as regulating the Internet. Accordingly, if we constrain people's rights and assume that the oligarchy "decides for the rest," Internet choices and practice, our ethics, morals, and understanding of the law, will reflect our actions.

Simplistically, if we adhere to the principles of rights as outlined in Table 5.5, then it cannot be tolerated that peaceful resistance against a government's technology or regulatory efforts should result in a person being watched or a security file being created. Our norms and values explain our moral standpoint, and understanding of social justice is and will be exemplified in today's and tomorrow's Internet.

The passing of the V-chip legislation in the United States and similar technology used to censor access to TV or Internet content raises the question of whether other countries will either use or legislate the use of such a chip. Canada has already decided that its use by manufacturers should be voluntary and depend on consumers' preference, that is, demand or lack thereof. Moreover, will the rating be able to distinguish and satisfy distinct cultural differences?

The Internet adds a new twist to the described issues of morality and ethics that are now becoming transnational (Okoshen, 1996). Accordingly, social conventions and knowledge, as well as norms and knowledge, impact our moral reasoning about using computer-mediated technology. To make the Internet a safe and productive place for users around the world, a better understanding about possible differences in moral reasoning across countries would be beneficial.

6 Morality, Privacy, and Codes of Conduct

OVERVIEW

This section provides the reader with some research findings about how people may feel about certain computer and Internet-mediated behaviors. An action that may be considered a personal matter by some may be perceived as immoral by others (passing on a computer game). Moreover, technologies are perceived differently when it comes to their possible intrusion of a person's privacy. Comparisons of code of conduct or ethics in computing for two different associations indicate that cultural and cross-national differences remain. This chapter suggests that the Internet is far from becoming an institution (Appendices E, F, & G).

Institutions are characterized by a social reality that is understood by all their members. Hence, members follow and adhere to similar, if not identical, codes of conduct addressing ethics and morals. This chapter shows, however, that differences exist according to culture, gender, and much more in how people may interpret the passing on of an illegal game via e-mail or in placing a virus on a BB. Here I present cognitive development theory that asserts that moral issues in all cultures involve justice, rights, or harm issues. Moreover, some have argued that individuals may not be able to identify the material and psychological consequences of their behavior on the Internet, possibly making them careless or even reckless.

To become an institution, Internet users would either have to adhere to the same ethical standards and morals and interpret certain behavior to be just or unjust or the majority would have to agree! However, even a comparison between codes of ethics among professional organizations having members in the computer field shows that differences exist in content, consequences due to violations, and ideals to be pursued with the help of these codes. Hence, differences and similarities between user groups on the Internet are plenty to be found in such areas affecting, for instance, privacy. Moreover, this could make it more difficult for governments to harmonize laws affecting privacy, security, and other matters on the Internet if their citizens interpret behaviors violating such rules or laws as being harmless. Hence, adhering to institutional norms and rules might be far from being realized due to some of the following issues. In fact, having the Internet evolve or develop toward a type of institution might simply be some politicians' pipe dream.

MORALITY AND INTERNET USERS

As pointed out earlier and in the previous chapters, culture, ethics, morality, justice, and rights are all important in how people interpret behaviors and how they act themselves when using the Internet and related technologies. In this section, morality is addressed; specifically, how it may pertain to people's cognitive interpretation of situations and their subsequent assessment or classification of others and their own behaviors.

Moral Issues and the Internet

As mentioned in chapter 5, morality is traditionally understood to involve an impartial constraint on a person's pursuit of interests. Moral issues are interpersonal and involve issues of harm, rights, or justice. Proponents of the cognitive development theory suggest that, by studying conventional or consensus-based obligation, we might be better able to determine how morals affect people's behavior.

Cognitive developmental theory was developed and tested by Piaget (1965) and Kohlberg (1969), and involves a series of stages. The theory postulates that conventional or consensus-based obligation is the origin of the idea of moral obligation, and obligation is not rooted in natural law (Turiel et al., 1987). The general skills of rational reasoning (e.g., deductive logic and the distancing of oneself from the consensus-based obligation) are associated with the idea of a moral obligation (Turiel et al., 1987). Progression through the stages of cognitive development depends on an individual's ability to develop a detached, impartial point of view to objectively evaluate a situation as either right or wrong (Turiel et al., 1987). Kohlberg indicated that the consequences of personal welfare are the primary focus of moral judgements (Haidt, Koller, & Dias, 1993).

Proponents of the cognitive development theory assert that moral issues in all cultures involve justice, rights, or harm; however, the rules of each culture vary (Turiel et al., 1987). The domain of morality is defined as "prescriptive judgements of justice, rights and welfare pertaining to how people relate to each other" (Turiel, 1983, p. 3). The domain theory of moral development postulates that the interpersonal consequences of events are used to place social events into three domains; personal, moral and conventional (Haidt et al., 1993).

Whereas ethics focus on overall values and beliefs (e.g., based on religion and culture), morals provide the individual with the necessary constraints to function in a society and with other Internet users. Shweder, Mahapatra, and Miller (1987) proposed that morals can be grouped into three domains that are discussed later (see also Table 6.1).

Personal Domain of Morality.   The *personal domain* is "outside the realm of societal regulation and moral concern" (Nucci, 1981, p. 114), and is based on personal preferences and tastes (Shweder et al., 1987). An example of a computer activity that could be categorized as residing in the personal domain is the use of encryption software for sending and receiving e-mail, because individuals want to keep their information secure and maintain their privacy. Accordingly, nobody is harmed in any way by people using encryption software to communicate with each other. Unfortunately, some national security agencies feel that they must be able to decipher such messages quickly or even at a later stage if so desired. If one believes these claims, unnecessary risks against national security or crime prevention are being taken by permitting citizens to use encryption technology to write each other a love letter.

Conventional Knowledge Domain of Morality.   The *domain of conventional knowledge* includes acts that have interpersonal consequences and are meaningful in a specific social system but are not harmful. Here, social norms, values, and attitudes play an important role in determining the meaning of a particular action. For example, designing a virus and distributing it to friends as a prank may be perfectly acceptable in one country, but may be objectionable in another country (Shweder et al., 1987). Likewise, depending on norms and values, using e-mail addresses or magazine subscription lists for advertising purposes may be tolerated in one country but not in another (Avrahami, 1996). To avoid legal and moral misunderstandings,

TABLE 6.1

A Social and Interactional Approach to the Domain Theory of Moral Development

	Moral Domain	Conventional Knowledge Domain	Personal Knowledge Domain
	Learned through direct observation of harm or injustice caused by a transgression	Learned through exposure to group consensus	Learned through exposure to others (e.g., during childhood) and past behaviors' outcomes
Material Conditions	Objective obligations: Justice, harm, rights, welfare, allocation of resources	Actions that are right or wrong by virtue of social consensus: Social uniformities and regularities, food, clothes, forms of address, sex roles	Psychological states, personal tastes and preferences
Formal Conditions	Rational, universal, unalterable, objective, self-constructed, more serious	Arbitrary, relative, alterable, consensus-based, socialized, less serious	Rational and irrational, arbitrary, relative, alterable, self-constructed
Description	Intrinsically harmful acts perceived directly, or inferred from direct perceptions	Acts that are not harmful, have interpersonal consequences, and are meaningful in a specific social context	The domain is outside the realm of societal regulation and moral concern
Infractions	1. Hitting another individual 2. Software piracy	1. Junk mail 2. Loading a computer virus program onto an electronic newsletter/listserver	1. Indecent acts 2. Use of encryption devices
Consequence	1. Social group may castigate 2. Legal or institutional (e.g., school—suspension, work—warning)	1. People may be puzzled or upset about behavior 2. Individual may be encouraged to change or face the consequences (e.g., social outcast)	1. Individual may feel uneasy or good about behavior 2. Based on input from reference group(s) or close friends/family, person may feel uneasy/good about behavior

Note. Adapted from Schweder et al. (1987, p. 24) and adopted by Gattiker, Janz, Greshake, Kelb, Schwenteck, and Holsten (1996). It is generally assumed that if something is considered harmful by the majority of a society, such behavior or action will be outlawed, thus resulting in legal sanctions if the person committing the action is caught. In contrast, a substantial minority of people may perceive something as being immoral, but such behavior may be quite common and in some cases accepted.

125

some Internet marketing firms are using opting in options. Here, the firm assumes that unless the subject provides them with explicit permission to send marketing material deemed of some interest to the user (i.e., opting in), the person opted not to receive such material (i.e., opting out). In the latter case, information about a possible free software upgrade for a product the client purchased earlier will not be sent.

Moral Domain of Morality.   Harmful acts, such as violence and theft, pertain to the *moral domain*. Intrinsic harm is perceived directly, or is inferred from direct perceptions (Turiel, 1983). Both children and adults reason that the act is universally wrong because the harm is intrinsic to the act (Haidt et al., 1993; Logan, Snarey, & Schrader, 1990), and, hence, the act is not tolerated. An example of a computer activity that is categorized as residing in the moral domain is a person who makes available a game that is considered to be illegal because of its violent and/or racist content.

Morality and the Internet.   Feinberg (1973) stated that people should be free to engage in harmless offenses in private even if those offenses breach a country's social or moral codes. The rationale is that ideas about the human construction of rules and laws, opinions about who and what should be included in the moral domain, and notions about private actions being free from external constraints are inherently personal. In contrast, Sproull and Kiesler (1991) indicated that users found the recognition process and the identification of potentially harmed individuals more abstract and difficult when computer technology was introduced into a situation. This implies that computer users may not be able to recognize an ethical dilemma because it is hard or even impossible to identify the material and psychological consequences to other users, individuals, and entities. However, I propose that users will recognize and identify unethical computer behavior with regularity because the utilization of computer technology has become a part of our everyday lives (similar to the telephone or calculator), and has spawned a new species of ethical issues (Johnson, 1989).

RESEARCH FINDINGS

As discussed in chapter 5, ethics focus on the overall values and beliefs (e.g., based on culture) whereas morals provide the individual with the necessary constraints to function within the Internet community. As outlined in Table 6.1, various behaviors can be grouped into three domains of morality, namely, personal, conventional, and moral. In a study investigating these issues (Gattiker & Kelley, 1995), it was reported that individuals distinguish between the personal domain (encryption device vignette), the conventional knowledge domain (virus vignette), and the moral domain (distribution of an illegal game abroad; see Appendix B for complete details on vignettes).

This does not support previous research results that reported that computer users may not be able to recognize and identify the material and psychological consequences of ethical dilemmas involving computer technology (Sproull & Kiesler, 1991). Older respondents are more likely to moralize their viewpoint, to feel that an act is wrong and harmful, or to be bothered by the action and to feel that interference is necessary (Gattiker & Kelley, 1999).

McClosky and Brill (1983) reported that responses for endorsing civil liberties drop significantly when respondents are asked about their perception of context-specific situations. Moreover, familiarity with the situation is also likely to affect people's assessment of a situation (cf. Strack & Föster, 1995). The following section discusses some research findings testing the three domains as outlined in Table 6.1.

Personal Knowledge Domain

Using an encryption scenario, Gattiker and Kelley (1995) found their data indicating that, whereas younger people felt it was okay to use encryption technology, older respondents felt it was bothersome, should not be permitted, and was harmful to others. Moreover, older respondents tended to moralize their standpoint. These findings suggest that younger people are more open to the use of privately designed approaches. Also, encryption software in the public domain is available on BBSs outside of the United States. Although it is illegal to ship such software via the Internet or the post office, it is perfectly alright to take a copy of the source code abroad and to offer the program again, thereby avoiding violation of any United States laws.

Based on these findings and the fact that younger people disproportionately influence the "Internet culture" (school and university students are on the Internet), enforcement of a standard technology may meet with substantial resistance from the younger generation. Hence, regulatory efforts as described in chapter 2 may be futile if a majority of users finds that violating them or taking advantage of technology being outlawed is all right. Even if certain laws limit the access to encryption technology outside the United States, people will obtain access to the necessary technology and software in order to pursue actions that they think are morally just in order to protect their privacy. Also, use of additional nonstandardized or nonsanctioned encryption devices by users is likely to be rampant if some laws try to restrict access to such software. Not permitting the selling of such technology may also encourage others to come up with software that is not subject to United States regulation. For instance, portions of the encryption key for Lotus Notes software was deposited with the appropriate United States government agencies before installing it for the Swedish Parliament. When this became public, Swedish media had a heyday and with the help of the Internet, many more people heard about it in early 1998. Therefore, one questions the practicality, enforceability, and usefulness of a legally sanctioned standard for encryption devices. This simply encourages the development of encryption software not being subject to United States regulations and, most importantly, hardly stops criminal elements from using the latest technology.

Besides practicality and usefulness of such regulation, it does, however, also challenge people's right to privacy in the United States and elsewhere if security agencies get access to message content through trusted third-party (TTP, see Appendix A) key depository systems. In contrast to telephone conversation that, after having been conducted, may never be recovered for analysis, any computer message can and might be recovered sometime in the future if the need arises. Hence, political maneuvering or a court case later in one's life may result in the revelation of personal information such as love letters sent more than a decade ago. Again, with a litigation culture in place, United States lawyers will have a field day if the technology is made available to reveal one's dirty or not so dirty secrets that occurred a long time ago in order to convince a jury in a civil lawsuit.

Conventional Knowledge

Gattiker, Kelb, et al. (1997) reported that people were most annoyed about the loss of privacy due to caller-number identification (CNID, see also Appendix D) being used to reveal somebody's phone number for marketing purposes. Similarly, people felt somewhat concerned about an organization using one's e-mail number for direct marketing purposes. But in both instances (Holsten et al., 1998; Gattiker, Kelb, et al., 1997), the major concern is that users were not most outraged by the vignette exhibiting behavior in the moral domain but instead in the conventional one. For the moral domain, harm is perceived and thus may violate rights and justice, thereby resulting in individuals (adults and children) rejecting such behavior (e.g., Haidt et al., 1993; Turiel, 1983). As far as communication-mediated technologies are concerned, however, this is not the case because e-mail tapping without the necessary warrant did not result in people assessing the behavior as wrong but they were more concerned about the CNID violation of privacy.

This indicates that how close to home an act might be (e.g., experience, familiarity) could very well affect people's assessment of moral issues substantially. Hence, asking people about their feelings or opinions of Internet content (e.g., pornographic material) or economic reform (Eurodollar replacing local currencies, the biggest hot potato for politicians and voters alike in EU member states during the latter 1990s) might be interesting. Unfortunately, this causes two difficulties: Respondents may start off with dissimilar amounts of knowledge about the subject and situation about which they are being asked to comment, and, not all subjects may have experience about the context matter investigated.

Accordingly, how people may respond to interviews from pollsters about certain Internet behaviors may not necessarily provide an accurate and thus, valid picture about Internet users' feelings, unless knowledge and experience levels are similar for all respondents.

Moral Domain

It is clear that viruses have become a nuisance for most people working with computers and nearly everybody has either experienced a virus on their hard drive or else knows somebody who has. Accordingly, as suggested by Table 6.1, harm experienced by the affected party has made the playing and passing on of viruses as being perceived morally unjust.

Using a vignette describing a person posting a virus on a Bulletin Board, Gattiker and Kelley (1999) reported that women were less permissive than men; they were more likely to be bothered by the virus scenario. Hence, women appeared more cautious regarding certain moral or immoral acts of computer users. The virus vignette provides the context link as suggested by McClosky and Brill (1983) and, in particular, attempts to address the familiarity issue. Most computer users have either had a personal experience with a virus, have tried to protect their machines and software from virus contamination, or have known end-users who have experienced the hassles and headaches of computer virus contamination. Furthermore, familiarity and acquaintance of the subject with the behavioral outcomes permits the assessment of the effect on moralizing. Therefore, computer users are probably more suspicious about the virus scenario because of their context-specific experiences.

Social Bonds and Skirting the Law

Literature based on sociology predicts that social bonds and ties to society (e.g., employment, active participation in community groups) may relate positively with people's moral development in each of the domains (Gattiker & Kelley, 1994, pp. 233–240). Unfortunately, findings (Gattiker & Kelley, 1995) have indicated that social ties do not have significant effects on the moral categorization of the three scenarios dealing with cyberspace issues. Moreover, Gattiker, Greshake, et al. (1997) reported that community size also did not affect people's assessment of direct marketing and e-mail tapping using the Internet, although based on literature, the researchers had predicted that respondents from smaller communities would be more cautious and would exhibit less moral tolerance than respondents from large urban centers. Neither of the previous studies support the premise that social bonds and small community size affect individuals' perceptions of morality. This is in contrast to literature indicating that ties affect a person's likelihood of becoming a delinquent (Matsueda & Heimer 1987).

Gattiker and Kelley (1999) reported that their data indicate that skirting the law is important for explaining the moral categorization of the game scenario: The higher the person scored on skirting the law, the higher the odds the respondent would categorize the passing on of an illegal game as morally acceptable (see also Gattiker & Kelley, 1995).

Cross-National Issues

The previous section indicates that people do not necessarily perceive behaviors the same way. Responses may differ based on social demographics such as gender and age, familiarity with the subject such as experience and time one has used technology, and the specificity of the context on which people are asked to reflect. On top of this comes cross-national differences. The Internet has no geographical boundaries and thus brings the world to every Internet user's home, regardless of whether he or she has been culturally sensitized toward differences in values and norms as well as rules adhered to by fellow users in different countries.

Gattiker and Kelley (1999) tested whether differences in people's responses as far as personal domain (encryption device vignette), conventional knowledge domain (virus vignette), and the moral domain (distribution of an illegal game abroad) could be obtained based on how United States respondents saw these issues in comparison to other countries. The findings indicated that American residency had a significant effect on people's moral categorization of the virus and the game vignette but not the encryption one. The effect indicates that American respondents are more likely than respondents from other countries to label placing a virus on an electronic network and the dissemination of an illegal game as appropriate or moral behaviors (Gattiker & Kelley, 1999).

Gattiker, Greshake, Schwenteck, Janz, Holsten, and Kelb (1997) did a study assessing differences in people's responses to conventional knowledge domain (CNID vignette and e-mail advertising scenario) and the moral domain (tapping of e-mail by security agency), comparing Canadian, German, and U.S. end-users. The findings revealed that whereas some gender differences exist in the United States (women are more cautious than men), they are limited in Canada and nonexistent for all practical purposes in Germany. Whereas some age differences were found with United States and German respondents, none were found with the Canadian sample.

Gattiker, Schwenteck, and Greshake (1997) reported that younger respondents in the United States and Germany were more tolerant than their older counterparts about privacy violations. The data also revealed that whereas Canadian respondents differed, based on the bother-action probe in the conventional domain (e-mail advertising), German respondents were concerned about the evaluation probe question in the moral domain (e-mail tapping by security agency). United States respondents were concerned about the harm probe of the conventional domain scenario (e-mail advertising). The universal probe in the moral domain scenario (e-mail tapping) led to statistically significant differences based on community size (smaller communities being less tolerant).

The Internet and Moral Development.   One explanation of the previous findings might be that although people read the code of conduct, an individual personal code of conduct might still differ (Pierce & Henry, 1996) and may thus be the deciding factor determining one's behavior. Another interpretation might be that subjects who are familiar with a code of conduct perceive a moral violation as unlikely or else as being okay; because the code is not violated, it must be alright. If the latter interpretation is right, the usefulness of these codes is questionable because they might encourage users to depend too much on them (cf. Gauthier, 1986; Gewirth, 1978).

Another concern that has not been addressed by research is what happens if the passing on of a virus is perceived to be a conventional knowledge issue in one country but a moral one in another. In the latter case, laws (e.g., Italy) may make the writing and passing on of a virus illegal and people could get prosecuted if caught. The research presented here indicates that younger individuals are far less concerned about the potential harm caused by a virus than are older respondents, and younger United States users are least likely to be bothered by a person passing on a virus program. Here again, moral constraints are influencing people differently in various countries and conflict is quite possible between various user groups.

The studies discussed here clearly indicate that users across cultures differ quite substantially in how they use their ethical frameworks to determine their moral constraints about behaviors on the Internet. Furthermore, differences can be found according to gender and age and, finally, the Internet community may in part develop different value systems and social bonds affecting people's moral and ethical judgments when using CMC technology.

PROFESSIONAL ASSOCIATIONS AND ETHICAL ISSUES: SIMILARITIES AND DIFFERENCES

The previous sections illustrate that ethics, morals, and justice are not necessarily shared by various user groups and people on the Internet from a variety of countries. Also, Fig. 1.2 suggests that the level of reach of norms and rules as well as the structure on the Internet differ between low and high. Accordingly, a code of conduct or ethics is likely easier to enforce on a local BBS or on an organization's own Intranet or LAN than on the Internet in general. As pointed out earlier (see also Table 5.2), professionals may adhere to the moral standards set by their employer, to their professional code of ethics, and to those of society-at-large. As Fig. 6.1 illustrates, the regulatory environment and the code's specificity as well as how much societal issues may be addressed could result in differences between the codes of conduct of associations within and across disciplines as well as countries.

The x axis in Fig. 6.1's cube assumes that we have two levels of regulation as far as privacy, safety, security, and confidentiality of data and information are concerned. The y axis outlines the level of specificity of the code of conduct. At the top, the degree of social justice is taken into consideration. It seems reasonable to want code of conduct to be grouped in Quadrants G and H but in some instances, specificity of a code may be low and/or social justice concerns are not addressed. Accordingly, a code of conduct may not assign rights and duties of members very well (see specificity) nor may an appropriate division of social advantages be outlined (e.g., social responsibility, see also Table 5.3). Finally, all this assumes, however, that the member follows the code of conduct as much as possible or otherwise faces some costs and/or consequences (e.g., loss of membership). Without

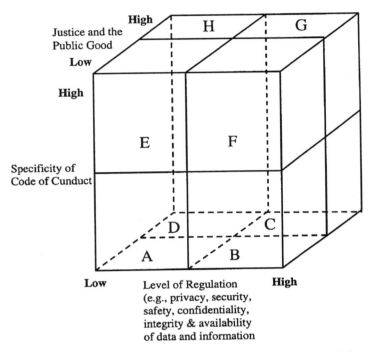

FIG. 6.1.   A model for comparing codes of conduct across associations. Regulatory environments may differ across profession (e.g., medical doctors vs. accountants) and natural countries. Specificity for members may be high by outlining exactly what is expected as well as the consequences in case of violation. Some codes may also have the social good or justice at heart by outlining social responsibility, volunteer and other possible areas perceived as more tangent to the member's primary professional duties.

the agreement by members to follow the code and live by its content, the code's use is very limited. In the following sections, this is addressed in more detail while specific codes are being discussed.

Professional Code of Conduct Across Associations in the Same Profession

Each profession's moral principles guide the conduct of its members. Morality arises from the application of the process of exchange. For instance, the individual benefit of neglecting to follow all steps of the code (e.g., not to copy software illegally) puts individual interests at odds with those of the association and all its members. The mutual benefit of an individual being a member in good standing requires an agreement between both parties. Although the association receives membership fees, it provides the individual member with various benefits (e.g., subscriptions, access to know-how, etc.). Later, I look at and compare two codes.

Association for Computing Machinery (ACM).   ACM (founded 1947) defines itself as

an international scientific and educational organization dedicated to advancing the art, science, engineering, and application of information technology, serving both professional and public interests by fostering the open interchange of information and by promoting the highest professional and ethical standards. (http://www.acm.org/about_acm/)

However, interestingly enough, one is hard pressed to find the ethical guidelines for members on the Web page. Although the latter is loaded with interesting information, the ethical guidelines cannot be found and sending the Webmaster a message in 1997 did not produce them either (as happened to me). The code deals with these sections; (a) general moral imperatives, (b) more specific professional responsibilities, (c) organizational leadership imperatives, and (d) compliance with the code.

The code also specifically words the content of each of these sections by starting with "As an ACM member I will...." These sections outline the member's general responsibilities as a member of ACM, as well as their professional (e.g., to keep abreast of new developments) and managerial (to assure ethical use of computers and respect of property rights) responsibilities. Finally, compliance assumes that the member complies with this code.

This code provides specific guidance for conduct, particularly situations that arise when working with computers and information systems.

German Computer Society.   When visiting the German Computer Society's [Gesellschaft für Informatik (GI)] Web page, the reader can go to the general information page and there, a link to the full text of the society's ethical guidelines (http://www.giev.de/uebersicht/ethische_leitlinien. html) is provided. Naturally, these are in German and the full text is not provided here. Nevertheless, these guidelines point out what the member's duties and responsibilities are (e.g., to be competent and assure that one is up-to-date with one's skills or else takes the necessary steps to assure this through continuous education). If one is in a management position, it is expected that the individual will follow the guidelines of ethics in a collective

sense (kollektive Ethik), that is, the individual will assess issues beyond his or her own interests and will assess issues from the collective's or society's point-of-view. Basically the code has four major areas; (a) the member, (b) the member holding a management/leadership position, (c) the member as educator and researcher, and (d) the German Computer Society.

In each of these areas, the responsibilities of members and the association toward society are outlined. Under point (a), the spotlight is primarily on the member keeping up-to-date through continuous education, technical, legal, and otherwise. What surprised me somewhat was that point (c) applies to educators only, but I would hope that all members are professionals and thus educate other end-users; therefore, all should serve as an example based on their behavior in their profession.

For a North American reader, the code appears very thorough at first but, nevertheless, very general because rarely is anything mentioned specifically. As such, no specific cases and examples are provided, as is the case for the ACM code. Also, although the code suggests obeying laws, it suggests neither a proactive stand nor an active contribution to the betterment of society (social responsibility). In a rapidly changing profession, however, one would assume that laws change only by having professionals pressure the appropriate groups into change (see also chap. 1 as well as Figure 5.1). Moreover, unless members are frequently confronted with the ethical guidelines, they are likely to forget their importance. For instance, we found that having ethical guidelines as an organization does affect ethical and moral attitudes by users, but only if they have read the code within the last 120 days; otherwise, forgetfulness sets in (Gattiker, Greshake, Schwenteck, Janz, Holsten, & Kelb, 1997).

Also interesting is section (d), which outlines succinctly what GI should do for its members as well as its responsibility as an association toward society.

Comparison of Codes of Conduct Across Professions and Countries

Kevin Bowyer (1995, chap. 3) compared the ethical code of conduct of the ACM with the ones by the American Psychological Association (APA) and the American Medical Association (AMA).

Bowyer (1995) pointed out three weaknesses in ACM's code, in comparison to the ones by APA and AMA, which are (a) implied limits to nondiscrimination, (b) duty to work to correct laws that are wrong, and (c) duty to be charitable.

Both APA and AMA clearly state that discrimination according to sexual preference is not acceptable, whereas ACM is not very specific (according to Bowyer, 1995, it is not clear what is meant by the wording of "other such factors."). ACM also lacks a clear statement about members being proactive in changing laws that are incorrect or simply unworkable. APA and AMA suggest that their members will have to do service without necessarily receiving remuneration for it for the betterment of society and the community in which they live.

The GI code does not address any of the three issues; this points out a cultural concern (see also chap. 4, e.g., Fig. 4.1 and Table 4.1). Accordingly, neither discrimination nor being proactive in the legislative process is really mentioned except that one should follow current law. But in a profession where technology and Internet-related development have outpaced the legal adjustments (see Fig. 5.1), this may likely result in some behavior that may be questionable from an ethical and a moral standpoint. Also, because voluntary organizations are somewhat less important in German society than in United States society, charitable contributions are not mentioned. However, "collective ethics" could be interpreted as including such work by members for the betterment of society. For instance, considering the needs of a voluntary organization and its likely lack of financial and human resources, it may cry for voluntary support by a GI or ACM member for updating its information systems or for simply computerizing some of its data bases. This would surely advance the collective good (kollektive Ethik).

Whereas ACM does mention privacy concerns, GI does not. But again, cultural differences may explain this. Whereas German law focuses on protecting data about an individual as held by a firm or agency (called *Datenschutz* = data security, i.e., one's responsibility to safeguard the misuse and abuse of data), the United States focuses on protecting an individual's privacy. Hence, the latter focuses on the human rights, whereas German law directs our attention toward protecting data and information but not necessarily a person's rights. For me personally, the latter will achieve the same as the U.S. focus on privacy, but more indirectly, and individuals are less likely to fight for something more abstract than a right, as efforts by United States privacy advocates and interest groups would suggest (cf. chap. 1).

Another interesting difference, particularly between ACM and GI (but also the Austrian Computer Society), is that ACM expects its members not only to abide by its Code of Ethics but, most importantly, to comply with it. It expects member to uphold and promote the principles of this code; and to treat violations of this code as inconsistent with membership in the ACM.

Nothing as strong can by found in GI's guidelines. This indicates differences between two associations' codes of ethics in two industrialized countries that, for all practical purposes, represent professionals in the same field. Even this short discussion should indicate to the reader that defining what is just and fair across national boundaries is difficult.

Do Codes of Conduct Make a Difference?

Although the codes of conduct of GI and ACM are similar in some respects, they are, however, quite different in many others as the above discussion illustrates (see also Fig. 7.1).

The ACM is very specific and expects each of its members to pursue adherence to the code and act proactively in their profession; the German's GI association's code is far more general and demands less adherence in following its standards than its U.S. counterpart. One may simply say that Germans tend to follow rules and laws more than their free-spirited U.S. colleagues; unfortunately, cultural differences may be far more complex and difficult to identify. For instance, Germans being more critical than Americans, differences in interpreting what work is and means as discussed in chapter 4 may in part explain some of the differences. If we assume that regulation levels as outlined on the x axis in Fig. 6.1 is the same for the two associations in their countries, social justice issues are not addressed very extensively in the GI code in comparison to the ACM code. Accordingly, whereas GI's code may justify a Quadrant F, ACM's may possibly be grouped in Quadrant G. Naturally, one may disagree and improvements in the code's social justice dimension are necessary based on comparisons with AMA's and APA's respective codes.

Nevertheless, the comparison of GI's and ACM's codes indicate differences and what is most important in our context is if codes do make a difference. Finally, do such differences between codes actually result in different behaviors? Molander (1987) reported that a well-written ethical code can increase the organization's commitment toward ethical conduct by its employees. Furthermore, a code outlines requirements and appropriate behavior to employees as well as the consequences one might face when violating these guidelines. Weisband and Reinig (1995) pointed out that management can influence employees' perceptions of privacy (e.g., the employees can be sure not to be monitored when the organization follows a hands-off-policy).

Kallman (1992) investigated the impacts of a formal code of conduct on employees' ethical behavior. Kallman reported that a formal code of conduct decreases the risk of unethical computer use. Pierce and Henry's (1996) study indicates that a formal code of ethics influences an individual's behavior by, for instance, decreasing the number of unethical actions. In their study of 356 information systems professionals, 30% of the respondents believed that the existence of a formal code of conduct/ethics is important for guiding the behavior of employees in making ethical decisions. Nevertheless, only 17% of the respondents use this formal code of ethics/conduct in making decisions.

Gattiker, Greshake, Schwenteck, Janz, Holsten and Kelb (1997) tested whether the existence of an organization's formal code of conduct about CIS and the use of the Internet positively influences the individual's moral categorization of privacy, that is, the person is more critical about other people's behavior when using the Internet (see Appendix F for complete details). The multivariate analysis of this hypothesis showed no significant result for any one of the probe questions. In the conventional domain (e-mail advertising), the interference-penalized probe provided a statistically significant result using a univariate test. Respondents who have no code of conduct in their organization tend to favor the penalization of the security agency, whereas people who have a code of conduct do not feel this way. This is a disconcerting finding because it suggests that people without a code of conduct are more worried about privacy than are others.

Assessing subsamples (i.e. Canada, Germany, and the United States), Gattiker, Greshake, Schwenteck, Janz, Holger, and Kelb (1997) reported that the data did not provide further statistically significant results either. Accordingly, the authors concluded that code of conduct does not appear to affect how people assess moral conduct in either scenario (i.e., conventional and moral domains) in any one of the three countries. More detailed investigation showed that only about 50% of the respondents read their firm's/provider's code of conduct recently and were familiar with it. Nevertheless, whether one read the code of conduct within the last 4 months or not was significant for the moral domain (tapping of e-mail & listserver communications) only, particularly in the evaluation probe. Surprisingly, people who did not read the code of conduct felt very "wrong."

Comparing the United States, Germany, and Canada, the authors also found that only in the United States do significant differences exist depending on whether the individuals read the code of conduct or not. Again, the differences were in the moral domain (tapping of e-mail & listserver communications), particularly in the evaluation probe and the interference-stopped probe. Respondents who recently read the code of conduct felt tapping of e-mail and listserver communications by a security agency was very wrong.

One explanation might be that although people read the code of conduct, an individual's personal code of conduct might still differ (Pierce & Henry, 1996) and might thus be the deciding factor determining one's behavior. Another interpretation might be that subjects who are familiar with a code of conduct perceive a moral violation as described in the earlier study to subjects, as unlikely or else as "being okay." Put differently, because the code is not violated specifically or clearly, the tapping of e-mail must be alright. If the latter interpretation is right, the usefulness of these codes is questionable because they might encourage users to depend too much on them (cf. Gauthier, 1986; Gewirth, 1978).

Based on the previous literature and research, the question must also be raised of how familiar members of the ACM and GI are with their society's respective codes of conduct. Quite likely, they have not read these for more than 4 months and thus might not be very familiar with them. Accordingly, violations may unintentionally occur, further indicating how difficult it will be to develop guidelines that help policymakers in regulating the Internet in a fair and just manner (cf. Organizations Push, March 8, 1997a).

Institutionalization and Code of Conduct

As previously discussed, the Internet seems far away from being an institution at this time, especially because not even professional associations such as the ACM and GI, having similar missions and interests, adhere to codes of conduct or ethics that are very similar. Instead, cultural differences and concerns may explain these differences (see also chap. 5).

Accordingly, the previous discussion suggests that although there is no unified ethical construct or social morality providing unified constraints on one's behavior when using information technology as a computer professional and member of ACM or GI, there are some similarities. Hence, social reality for members in these two societies is quite similar but not the same, resulting in conflicts between members if they work for the same organization but in a different country (i.e., the United States or Germany). For the Internet user at large, the story may not look as promising because no code of conduct can be found. Norms and rules acknowledged by users around the world have been evolving and still are. As Fig. 1.2 and Table 1.1 suggest, however, differences for various user groups exist.

Morality and the "MTV" Generation.   The previous discussion indicates that cognitive and attitudinal antecedents, as far as cyberspace is concerned (e.g., ethics and morals), will influence our behavior when surfing the Net. This will be greatly leveraged by the younger generation, whose participation rate on the Internet is already greater than the one for the older generation. In the past, we have experienced that younger people may participate in subcultures (e.g., during the 1960s) and their tastes for music are probably somewhat different than for other groups (e.g., techno music). Nevertheless, aging may accelerate the process of integration of these subcultures into the more mainstream of society. Hence, what these subcultures (e.g., Hamburg's Chaos Club—computer hackers and net surfers as well as phone phreakers) do today will affect us tomorrow. Whereas the 1960s generation may have had a somewhat unsatisfactory influence on a country's institution by simply being integrated or absorbed into the system, today's MTV Internet user is influencing the system from within. As such, young people are often the most sophisticated user group on the Internet and with their actions, they shape its use, opportunities, and threats. Again, regulation and cheap access to the Internet in some countries as well as cultural differences have resulted in North American youths being a very influential and diverse group of users on the Internet (e.g., newsgroups, listservers, personal web pages and content as well as chat forums). Their values and interests may have an impact reaching far beyond their size and location. Only the future will tell us how this might have influenced the institutional norms and rules adhered to by many using the Internet in a few years.

SUMMARY AND CONCLUSION

This chapter further indicates that using the cognitive development theory, differences in how users may interpret Internet-related behaviors can be found. Specifically, using demographics and comparing users across countries results in differences in how people feel about Internet behaviors (e.g., tapping e-mail or sending advertising via e-mail). What is especially important is that younger people appear far more relaxed as far as viruses, illegal computer games, and using encryption technology is concerned. Moreover, their feelings toward security agencies tapping e-mail communication indicates that they perceive such action as an infringement of their rights and see such behavior as immoral and unjust. The regulatory developments outlined in chapter 2 and by Gattiker, Fahs, Blaha and EICAR WG1 (1998) suggest, however, that security interests by the defense establishments in various countries would like to curtail privacy and people's rights to use highly sophisticated technology to protect their data and information. Findings presented here suggest that younger Internet users might violate such laws in large numbers and thus making their enforceability difficult. Most importantly, curtailment of such rights should probably be questioned anyway.

Furthermore, although codes of conduct might help users to follow the norms and reduce possible violations of others' rights, unless the user is aware of the code, the likelihood of adhering to it is reduced significantly if one has not studied the code for more than 3 months. Codes of conduct for information systems and Internet use do exist (see also Appendices F and G for sample codes), and this chapter shows that the differences are substantial and primarily due to cultural differences (see also chap. 5). Moreover, rules, norms, and behaviors on the Internet are still evolving, and continue to evolve, and are driven by various user groups and legislators.

There are two final concerns that should be raised. Codes of conduct or moral constraints for that matter may be violated because an individual's personal code of conduct differs (Pierce & Henry, 1996). Moreover, a subject might feel it is okay to conduct a certain behavior simply because it is not specifically mentioned in a code to be inappropriate. Accordingly, with this reasoning and the findings reported in the studies mentioned here, it is, unfortunately, unlikely that codes of conduct by associations and organizations might help reduce the potential for conflict between user groups with different demographics (e.g., age and profession) and from different countries. Moreover, discussing these codes of conduct indicates that we are wrong in assuming that moral guidelines that supposedly put some constraint on our behavior are universal across cultural boundaries. The issues addressed here illustrate that there is much to do but little time—considering the rapid change we face and how much the Internet is affecting both our work and nonwork lives.

III | Conclusions and Implications: Where Might It All Lead?

The last part of this book tries to further outline some of the issues discussed previously by going into more depth. Particularly, here we are interested in finding out more about how regulatory, economic, cultural, and ethical issues may influence the further evolution and development of CMC and of the Internet in particular. Chapter 7 focuses on marketing and e-commerce issues. It uses stakeholder theory to present a framework for better understanding various facets of interaction between Internet user groups such as the firm offering information on its Web site and the site visitor. Discussing socialization, cognitive and situational factors, various issues of concern to Internet users are outlined while a framework for better understanding Web and Internet behavior is presented. The latter is used as a basis for outlining some of the research questions that need to be addressed to increase our understanding about this new medium of communication for noncommercial and commercial purposes.

Chapter 8 is the conclusion of this book. It tries to summarize some of the issues outlined earlier and also introduces a schemata for better understanding virtual communities. In particular, it points out that virtual communities may have some distinct disadvantages compared to social communities. Moreover, the chapter outlines where government and business as well as not-for-profit organizations may face some challenges in the near future if they intend to take the best advantage of new technology.

7 Stakeholders and E-Commerce

OVERVIEW

Business use of the Internet as a new tool for marketing its products is continuing. Although media is focusing on retail issues, business-to-business marketing and its opportunities have gathered considerable interest from the business community. Nevertheless, our understanding of e-commerce marketing and the Internet based on scientific theories and empirical investigations is still scarce. Unfortunately, knowledge and insights from other media such as television and telephone (e.g., advertising and sales) might have limited applicability to the new Internet medium. A conceptual framework and the research questions permitting the testing of its applicability are outlined here. Practical implications are discussed (Appendices B & H).

Previous chapters in this book outline telecommunication policy, regulation, Internet access costs, ethical and moral concerns experienced by Internet users from around the world. In particular, I focus on how regulation and economic issues may affect Internet use and access. Also, how codes of conduct might possibly affect people's consideration of possible effects on others based on their behaviors was outlined. Nevertheless, comparing codes of conduct indicates that they differ in their focus and are far from providing a unified picture within a profession across nations. In this chapter, I discuss e-commerce issues as they pertain to previous chapters and outline the lack of knowledge and understanding we still have as far as this new medium is concerned.

Chapter 2 outlines three issues raised by market regulation (why regulation, should government intervene, and if so, how). Based on the discussions and issues addressed in previous chapters, some research questions and hypotheses are put forward here. The chapter presents a model on how three dimensions, namely cultural, type of web user, and supplier of information can be used in better classifying important Internet stakeholders for the firm. Socialization, cognitive and situational factors are discussed that might influence the use of Web technology by various stakeholder groups. A model outlining independent, moderating, and dependent variables of web user behavior is outlined and discussed. This chapter suggests that although e-commerce and the use of the Internet is expanding rapidly, our understanding of this new technology is somewhat limited and will probably continue to lag behind the rapid developments we are experiencing. Nevertheless, research is needed to better understand Web and Internet user behavior, which, in turn, will make regulation and deregulation concerning safety, security, and confidentiality of data as well as privacy more effective.

LOOKING AT STAKEHOLDERS

In previous chapters, the discussions have encompassed organizational and private Internet users. Increasingly attention has been on using Web technology for commercial purposes such as sales or as a communication channel between firms and their customers (retail and other businesses). For using the Web, the user takes advantage of a graphical software browser and usually remains online while "surfing" (see Appendix A for a definition of the term Web) and downloading materials (e.g., pictures and product information). How the stakeholder concept might be applied to the Internet and using the Web to gather information is presented here. A framework for classifying important stakeholder groups is outlined.

Stakeholders and the Organization's Web Site

A firm's visitors to its Web site may be looked at as stakeholders. Stakeholder theory (Donaldson & Preston, 1995; Freeman, 1984; Ulhøi, 1997) and experiences from environmental stakeholder pressure (Ulhøi & Madsen, 1998) suggest that a thoroughly performed stakeholder analysis can improve the likelihood of optimizing corporate strategic initiatives. For a long time, researchers have pointed out that stakeholders' importance for managing will increase (e.g., March & Simon, 1958) and recent findings indicate that satisfying their needs continues to be a paramount issue.

Mitchell, Agle, and Wood (1997) suggested that we expand our theoretical concept of stakeholder typology and distinguish stakeholders according to their power, legitimacy, and urgency while grouping them into latent, expectant, and definite stakeholders. In this chapter, I focus on expectant and definite stakeholders for the firm and its Web activities. Mitchell, Agle, and Wood (1997) suggested that expectant stakeholders are made up of dominant stakeholders (e.g., large institutional investors, employees, key customers), dangerous stakeholders (e.g., wildcat strikers, terrorists), and dependent stakeholders (e.g., public demanding that firm reduces pollution may unite with media or government to get action). Definite stakeholders have power and legitimacy already, hence a large customer visiting a firm's Web site may force it to change its pricing policy by offering lower prices on the Internet than if an order is called in on the toll-free/free call line.

Figure 7.1 illustrates that we can further distinguish between stakeholders according to three dimensions, namely, suppliers, customers, and users from different geographical markets visiting one's web site. Naturally, other important stakeholder groups are investors and employees. Thus, Fig. 7.1 could include those groups instead as well. I am focusing on suppliers, customers, and users from different geographical markets for simplicity's sake. Naturally, for realizing potential benefits through just-in-time manufacturing as well as e-commerce opportunities with the Web, suppliers and customers are important stakeholders. Additionally, cultural differences must be considered because the Web removes geographical boundaries, enabling people from afar and with different cultural backgrounds to visit a firm's Web site.

It is also important to point out that stakeholders represent a continuum ranging from latent, expectant to definite stakeholders. Most importantly, a stakeholder group's position on this continuum is in a state of flux. Accordingly, whereas media or the public may be latent stakeholders about one's product today, tomorrow they may become a definite one! To illustrate, one of the world's most widely used e-mail programs, Lotus Notes, is not so secure as most of its 400,000 to 500,000 Swedish users believed in late 1997. To be sure, it includes advanced cryptography in its e-mail function, but the codes that protect the encryption have been surrendered to American authorities. With them, the U.S. government can decode encrypted information. Among Swedish users in December 1997 were 349 parliament members, 15,000 tax agency employees, as well as employees in large businesses and in the defense department (Laurin & Froste, 1997).

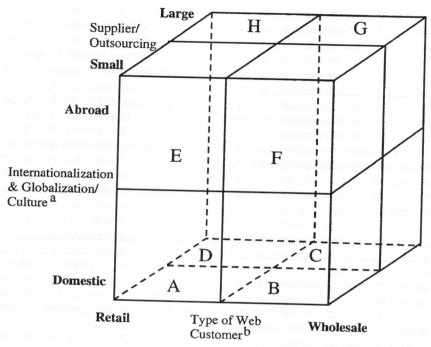

FIG. 7.1. A model for classifying some important definite stakeholders in domestic and other markets. Most stakeholders are customers as well as suppliers; that is, the relationship has to be satisfying for investors (customers) in the form of high enough annual/quarterly dividends to assure their continuance as supplier of capital to the firm. If product quality and price are not satisfactory for customers, they will withdraw their support as supplier of funds by spending their money for products and services provided by a competitor instead.

[a] A supplier could be an investor or a government agency providing a firm with certain services (e.g., fair taxation system). Moreover, a supplier represents an opportunity for the firm to provide important information affecting its business. For instance, new materials used in components manufactured by the supplier might further improve a product's characteristics such as its heat resistance. In turn, a supplier can also be a firm's customer by taking advantage of information supplied on the Web site. To illustrate, the firm's WWW site might provide important information online about new products or industry developments as well as regulations affecting suppliers.

[b] Customers are also suppliers in so far as they provide a firm with input in the form of information (e.g., product quality and reputation in comparison to competing products) as well as cash by purchasing products. Thereby, customers are permitting the organization to learn from their needs, expectations, and expertise (Ulhøi, 1997).

When discovering these facts, Swedes felt that privacy and security of communication between parliamentarians and various government agencies and Swedish firms had been compromised. A latent stakeholder (Swedish government) suddenly became an expectant one that wanted this problem taken care, while IBM had to also deal with U.S. government parties that wanted nothing changed. Moreover, a latent stakeholder (e.g., media and information security professionals in Europe) suddenly became a demanding group of definite stakeholders damaging Lotus Notes' credibility as being secure for electronic communication. Hence, we can label stakeholders according to latent, expectant, and definite ones. However, a particular group of stakeholders (e.g., media or customers) may change its position on this continuum (i.e., move from being a latent to a definite stakeholder or vice versa) as the previous example illustrates.

The Three Dimensions of Fig 7.1.   Figure 7.1 shows that for customers, we distinguish between retail and wholesale customers (see x axis of cube). The supplier dimension (or y axis of the cube) represents the internationalization and globalization factor by distinguishing between international and domestic customers. The final dimension in the cube distinguishes between small and large suppliers for the firm. The cube assumes that a customer (e.g., purchasing agent) may also be a supplier (e.g., of information such as type of products purchased at what date). A supplier may also be a customer. For instance, the supplier firm may be looking to be provided with the appropriate specifications to start designing its components needed for the production run of a new product by the purchasing firm.

The x axis in Fig. 7.1's cube assumes that an individual may be using a different set or hierarchy of criteria for making the purchasing decision (e.g., price is more important than features of product) than the corporate purchasing agent buying for the firm. Consumers are significantly influenced by information from credible sources in making purchasing decisions (e.g., Rosen & Olshavsky, 1987), whereas corporate purchasing agents focus on ordering procedures, delivery, and product information (e.g., technical statistics) provided about the product (Ravi, 1997).

Figure 7.1 suggests that whereas domestic customers (retail and wholesale) require pertinent information in their local language(s), foreign clients may require everything in English. The latter may not be the person's mother tongue; hence, structuring of information and content has to be such as to minimize misunderstandings. Accordingly, because humor is very much influenced by culture (e.g., popular TV series), it could be misunderstood by some groups of Web site visitors from abroad (Berger, 1997; Bergmann, 1987; Gattiker & Hedehus, 1999).

Suppliers could be differentiated according to size (e.g., contract work that has been outsourced to one individual doing telework from home; Düerrenberger, Jaeger, Bieri, & Dahinden, 1995). Large suppliers may be providing the firm with materials or components needed regularly to produce its products (e.g., components delivered daily as part of a just-in-time manufacturing system; Wildemann, 1997).

As far as e-commerce and shopping on the Web is concerned, however, our knowledge is limited on how consumers may evaluate brands and products and their characteristics before making a decision about a purchase using the Web versus shopping at a store. Why are these issues important? For instance, in 1998, it was reported that 25% of corporate purchasing agents were planning to increase use of the Internet for buying industrial supplies (Treese, January 26, 1998). Moreover, a survey with U.S. respondents reported that nearly half of them indicated that they may use the Internet to purchase their next car in 1998 compared to 32% in 1997 (Nissen, 1998). These findings indicate that in some industries, expectant and definite stakeholders are changing their behaviors and intend to use Web technology in certain ways for accomplishing particular objectives. Accordingly, researchers and organizations should respond to these developments in a proactive fashion.

In summary, Fig. 7.1 outlines that a firm's Web page and related activities are being assessed by expectant and definite stakeholders who may evaluate the firm's performance by considering three aspects; namely, the type of Web customer one might be (e.g., retail vs. wholesale), domestic or international, and supplier or outsourcing perspective. Accordingly, a hostile stakeholder may become a definite threat by putting obscene material on the firm's Web page upsetting suppliers and customers from abroad unless the person is stopped. Also, a firm is either unable or very likely unwilling to restrict access to its Internet Web page because this would defeat the purpose of reaching as many parties as possible; hence, hostile visitors and clients may all enjoy or reject content. The schemata outlined in Fig. 7.1 should help in better categorizing various issues to be considered when managing a firm's cyberspace activities as outlined here. We need to understand better how the issues addressed here and graphically outlined in Fig. 7.1 might affect a firm's Internet activities and electronic commerce in particular. How different stakeholder groups based on different measures might perceive the Web opportunities differently and thereby take advantage of services in possibly novel ways is important.

E-COMMERCE AND MARKETING

If we look at expectant and definite stakeholders, while grouping them according to type of customer, globalization, and size of supplier dimensions as suggested in Fig. 7.1, we need to determine how we can learn more about these groups' needs, attitudes, and behaviors when visiting the firm's Web page. Table 7.1 outlines the socialization, cognitive, and situational factors in schematic form as related to the use of Web technology. The following sections review research dealing with these issues by discussing how socialization, cognitive, and situational factors may influence users' Web behavior (cf. Table 7.1).

Socialization Factors

Social science researchers use the term socialization to describe the process by which individuals learn the "script" on how to behave according to rules and norms in a society. The following discussion is organized around (a) sociodemographic factors, (b) cultural variables, and (c) community size.

Sociodemographic Factors.   A significant body of research has examined the effects of the aging process on attitudes and motivation of employees. For instance, Rhodes (1983) suggested that older workers often have higher work-related motivation matching their higher satisfaction. Other research has reported that older subjects are more concerned with corporate ethics than are their younger counterparts (e.g., Thumin, Johnson, Kuehl, & Jiang, 1995) and are more concerned about moral violations on the Internet than younger computer users (Gattiker & Kelley, 1999).

TABLE 7.1
Variables and Use of Web Technology

Socialization Factors	Cognitive Factors	Situational Factors
Sociodemographic variables	Information processing and judgment	Economic variables
Cultural factors	Privacy and safety/security	Other factors – technology resistance – attitudes – type of technology user

Psychological and sociological literature has reported many differences between genders; for instance, women have different attitudes toward computers and TV watching than men (e.g., Gutek & Larwood, 1987; Newton & Buck, 1985). Moreover, women are more concerned about people's behavior on the Internet than are men (Gattiker & Kelley, 1995) and using a college "freshman" sample, Wilder, Mackie, and Cooper (1985) reported a "growing dislike of the computer due in large part to women's negative attitudes toward the technology" (p. 223).

Cultural Variables.   A large body of research has also addressed cultural differences. As outlined in chapter 4, Hulin, Drasgow, and Komocar (1982) reported different results when asking bilingual subjects (Spanish and English) to fill out surveys in two languages (i.e., with 2 weeks interval in between filling out the surveys) on the same topic. The authors concluded that even translating an instrument as accurately as possible (e.g., using back translation to assure content validity) may still not assure cultural equivalence of the instrument used.

Other research has reported that users in various countries do not have similar technology attitudes and are sometimes concerned about different side effects than their colleagues in neighboring countries (e.g., Gattiker & Nelligan, 1988). Becker and Fritzsche (1987) pointed out that U.S. managers were noticeably more concerned about business ethics than their French and German colleagues. But cross-national differences also affect responses to surveys. For instance, Chen, Lee, and Stevenson (1995) reported that when comparing Japanese, Taiwanese, Canadian, and U.S. high school students on a scale assessing individualism and collectivism, the Japanese and Taiwanese students were more likely to use midpoint on the scales than their North American counterparts. High school subjects from the United States were most likely to use extreme values.

Although the Internet has removed geographical boundaries, people in various countries may interpret and perceive privacy issues differently. For instance, Gattiker, Schwenteck, et al. (1997) reported that German respondents were more likely to trust government regulation on privacy matters in regard to e-mail advertising, whereas U.S. subjects were more suspicious and ready to defend their rights.

Community Size.   Besides cross-cultural differences, empirical work has also indicated that community size is linked to crime rates. For instance, Sadalla (1978) reported that crime rates are higher in urban areas than in rural ones. Mayhew and Levinger (1976) tried to explain this phenomenon through decreased interaction time and increased anonymity in cities. Moreover, larger communities are more tolerant toward unconventional sexual behaviors (e.g., Wilson, 1995) and exhibit higher crime rates due in part to reduced cohesion and reduced feelings of community as compared to smaller communities (e.g., Alba, Logan, & Bellair, 1994).

Some early research dealing with the Internet reported, however, that community size has little, if any, effect on values and behavior on the Net but suggests that the Internet community may, instead, represent a community with its own sets of values (e.g., Gattiker, Greshake, et al., 1997).

Cognitive Factors

Social scientists generally agree that socialization influences and interacts with mental processes such as perception, reasoning, judgment, and decision making. These cognitive processes are now addressed.

Information Processing and Judgment.   Decision makers persist in taking risks if prior risk-related actions were successful (Sitkin & Weingart, 1995), although based on prospect theory (Kahneman & Tversky, 1979), positively framed situations lead to the making of risk-averse decisions. Venkataramani, Kamel, and Jacoby (1997) reported that consumer evaluations of new brands result in information seen later online being given a greater weight for making a decision than information seen earlier. Moreover, when people are permitted to make monetary trade-offs among available alternatives of choices, the effect of attractiveness difference on choice deferral decreases significantly (Lawson, 1997).

In the context of the Internet, one could believe that if prior shopping experiences were successful (i.e., taking a risk by ordering the product online; Sitkin & Weingart, 1995) people might be willing to shop again. Research has also indicated that consumers select brand name and price most frequently from larger information arrays when acquiring information in brand choice situations (Jacoby, Szybillo, & Busato-Schach, 1977).

Extrinsic and Intrinsic Product Attributes.   Marketing literature has used the economics of information approach (Nelson, 1970), which suggests that people purchase both *search goods* (quality can be evaluated before purchase) and *experience goods* (quality can be evaluated after purchase only). A multidimensional notion can be added, because products include both search and experience characteristics (Wilde, 1980) as well credence characteristics (question of credibility of the seller in the eyes of the buyer; Andersen, 1994).

According to Grunert (1997), these three characteristics have been incorporated into multiattribute models. In turn, this permits distinguishing between extrinsic attributes (such as brand, price, and sales outlet or channel) and intrinsic attributes (attributes of the physical product such as product features, design, looks, and taste if it is edible). Literature also indicates that for food items, people use visual cues to select the product (e.g., how appealing does the meat look; Grunert, 1997). For more technology-related products (e.g., mobile/cellular phone or pager) individuals may focus more on such features as price and brand recognition (e.g., Smith, 1996).

Privacy and Security.   In addition to how people process information and use extrinsic and intrinsic attributes to evaluate a product offered for sale, privacy may be a concern for Internet shoppers. For instance, a Harris poll indicated that 85% of Americans are concerned about their privacy (Harper, 1993). Stone and Stone (1990) defined privacy as the individual's ability to control information about her or him to isolate him or herself from unwanted auditory (see also Appendix A for a definition of privacy). Consequently, the individual has to be able to regulate the amount and nature of social interaction. In the context here, I suggest a definition that is based on the individual's right to determine his or her own communication contacts and the right to control the use of personal information by others (Gattiker, Kelb, et al., 1997, p. 606). Additionally, I propose that it should be made technically and economically feasible for the individual and for commercial organizations to control and protect their own private data to an extent that they themselves determine and, as importantly, with measures selected at their own discretion

How individuals may feel about privacy with communication technologies and the Internet in particular could also affect how extensive security and safety measures should be to make people feel comfortable. Research has indicated that individuals do assess privacy intrusions through a phone call (e.g., somebody trying to make a sales pitch) as less severe than receiving spam mail (advertising through e-mail; Gattiker, Schwenteck, et al., 1997).

Most people believe that they have already lost the right to control their personal information, because new technologies and large databases facilitate the increasingly effective use of information about consumers and citizens (Schlossberg, 1993). Total protection of privacy is impossible, and even taking a violator to court can be difficult and expensive (van Swaay, 1995). Nevertheless, unless people's concerns about privacy matters can be addressed to their satisfaction, e-commerce will be hampered severely.

Situational Factors

Socialization and mental processes affect how people take advantage of WWW technology. Literature suggests that while social conditions are important for explaining behavior, they may also create climates that tolerate certain behaviors (American Psychological Society, October 1997, p. 15). Accordingly, situational factors contribute to people's behavior as follows.

Economic Variables.   Here, the term includes costs for Internet and thus Web access through telecommunication charges, Internet access charges as well as income as discussed in chapter 3. Some research indicates that telecommunication charges do affect the volume of traffic as well as the origination of telephone calls. Accordingly, more calls originate in the United States for Europe or Asia than the other way around, whereas the country with the lower rates usually originates more calls between two countries than the one with the higher rates (Donzé, 1993).

In addition to these costs, various comparisons point out that users in different countries do pay rather high Internet access fees as well as telephone costs. A user may not have to pay extra for each additional minute of surfing to the telephone or cable or to the Internet service provider for accessing the Web (see also Table 3.1).

Finally, income has been used in many studies to assess how people behave. For instance, shopping patterns of individuals differ based on income and consumers in richer countries spend more on soft drinks than in other countries. Moreover, TV watching patterns also differ in that higher income groups tend to watch less TV and more news and investigative reporting types of programs than other income groups (cf. McCarty & Shrum, 1993).

Beliefs and Other Situational Factors.   In addition to economic factors influencing people's behavior on the Internet, so may their beliefs and situational factors (e.g., about privacy invasion). Beliefs are defined as concepts and perceived relationships between things and concepts that individuals hold to be true (Bem, 1970). An individual may have certain beliefs about a technology. Although the amount of time having used a technology may not result in differences in beliefs or attitudes toward a technology, gender has been reported to result in women being more concerned about computers' potential of control than men (e.g., Gattiker, Gutek, & Berger, 1988).

Using various related communication technologies may also affect a person's possible acceptance of technology in one's life. For instance, cellular telephones are perceived as a more masculine device and women are more likely to use the mobile phone for personal or domestic reasons (Rakow & Navarro, 1993). Here we assume that a person using a cellular phone or a digital camera could have more positive and accepting beliefs about the Web than others might; however, research addressing this seems lacking.

Besides beliefs, research has reported situational influences on preferences for purchasing consumer products, namely the purchasing situation and the purchasing target (e.g., self vs. friend). Schmitt and Shultz (1995) reported that people evaluate colognes differently depending on who the purchase target might be and if they can view, smell, or view and smell the colognes. It might be that if neither testing nor sampling is possible on the Web, people could use their friends' opinion and possible experience with the product as a more important choice factor before purchasing. Schmitt and Shultz (1995) reported that positive prior experience did affect people's choice of preferred cologne in their study.

WEB USER BEHAVIOR: A CONCEPTUAL FRAMEWORK

In the previous section, I addressed how socialization, cognitive, and situational factors have been used for explaining people's behaviors, attitudes, and acceptance of new technology (cf. Table 7.1). Moreover, I also proposed that applying the stakeholder approach would allow us to better capture how various stakeholder groups may believe in and feel about as well as use a technology differently than another group (cf. Fig. 7.1). Accordingly, a large supplier or a business customer may evaluate intrinsic or extrinsic product attributes differently than a small supplier and/or client while cultural factors may also play a role, such as U.S. customers being more concerned about privacy and e-mail spamming than clients from other countries.

What distinguishes the approach of this chapter from earlier work in this area is that literature from diverse fields and disciplines was used to develop and discuss the framework as outlined in Fig. 7.1. Such a comprehensive approach will benefit the field of Web and Internet technology by advancing understanding and knowledge about the processes and variables involved in the use of the Web.

In the following section, research is discussed that has specifically applied portions of the framework presented earlier to the Internet and the Web in particular. This approach will help identify the gaps in past Web research as I assess the framework developed earlier in this chapter.

Identifying the gaps in the Web literature will then enable us to develop propositions to be tested in future research. Such testing is needed to determine to what extent, if any, the framework developed here provides a better, more consistent framework for understanding Web users' behavior when surfing. In the area of Internet and WWW technology, the time is ripe for developing and testing a comprehensive framework and for comparative testing of predictions derived from the model. The following propositions are, of course, not all inclusive but limit themselves to areas where research is very scarce at the moment and, therefore, is needed to advance knowledge in the Internet domain of Web surfing.

Figure 7.2 schematically illustrates how the interrelationships between various factors influence a person's behavior on the WWW (e.g., how long and often one surfs). The following sections review research dealing with these issues by discussing how socialization, cognitive, and moderating situational factors relate to the outcomes of Web behavior; namely, how people spend their time on the WWW, possibly shopping or purchasing behavior and their attitudes toward this technology.

Economic Factors and Information Processing

This section reviews the literature according to Table 7.1 and as presented graphically in Figure 7.2. Moreover, some research propositions are presented based on the reviewed literature. These propositions point out that much work still needs to be done before we have a better grasp of the issues and factors affecting users Internet and Web behavior in particular. In other words, before we have come up with some answers and explanations for people's Web behavior we need some answers to the propositions presented below.

Extrinsic and Intrinsic Product Features and Pricing.   The literature has indicated that extrinsic attributes such as price and product warranty, as well as brand and store name play an important role for consumers' quality judgments and possible purchase of the product (e.g., Chang & Wildt, 1996). Accordingly, a comparison of how people evaluate products would require looking at similarly priced products such as cellular/mobile phones, computer scanners, and digital cameras costing around U.S. $300.00. But what is also of interest here is whether there is a difference in the use of extrinsic attributes in deciding to purchase a product for approximately U.S. $100 versus deciding to purchase a product for U.S. $1,000.

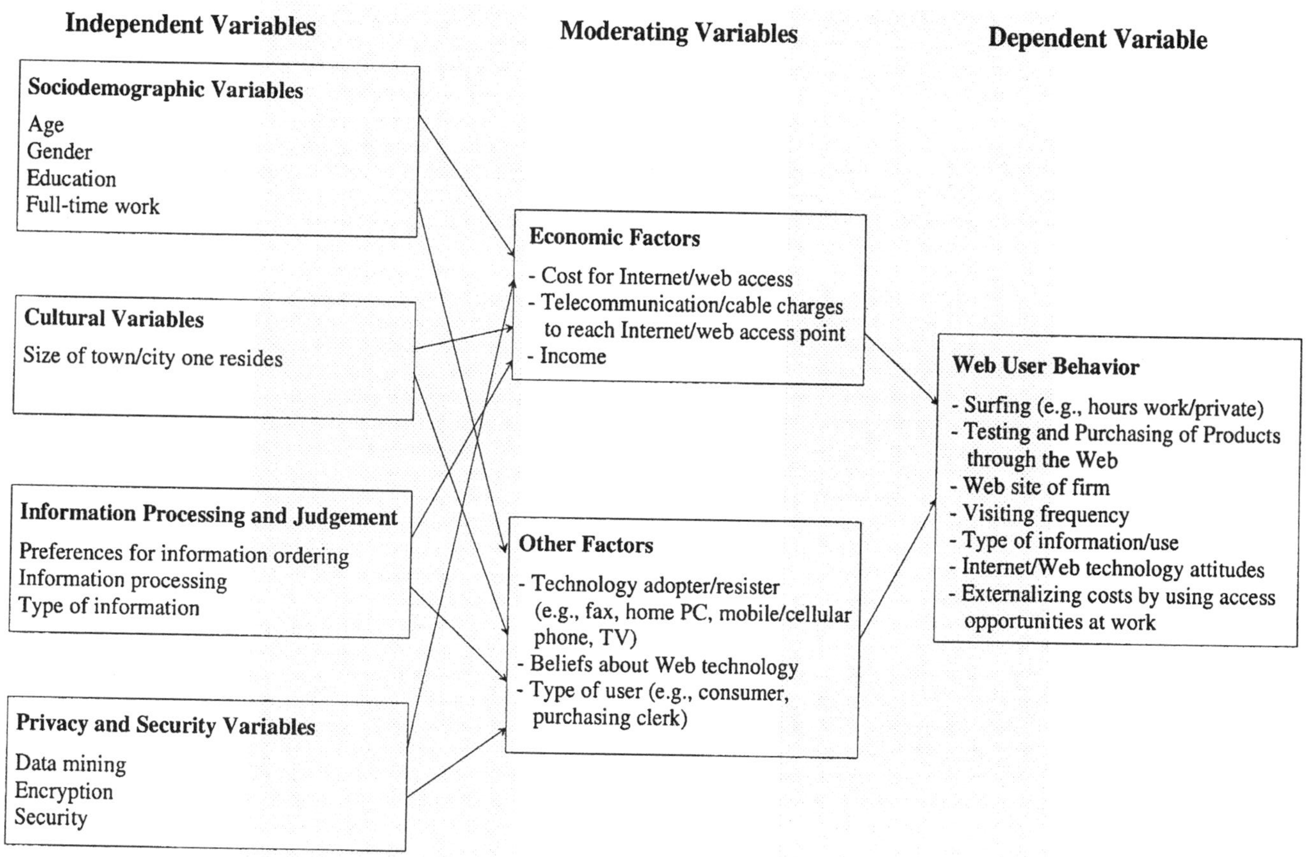

FIG. 7.2. A framework for understanding Web behavior.

Independent Variables

Sociodemographic Variables

Age
Gender
Education
Full-time work

Cultural Variables

Size of town/city one resides

Information Processing and Judgement

Preferences for information ordering
Information processing
Type of information

Privacy and Security Variables

Data mining
Encryption
Security

Moderating Variables

Economic Factors

- Cost for Internet/web access
- Telecommunication/cable charges to reach Internet/web access point
- Income

Other Factors

- Technology adopter/resister (e.g., fax, home PC, mobile/cellular phone, TV)
- Beliefs about Web technology
- Type of user (e.g., consumer, purchasing clerk)

Dependent Variable

Web User Behavior

- Surfing (e.g., hours work/private)
- Testing and Purchasing of Products through the Web
- Web site of firm
- Visiting frequency
- Type of information/use
- Internet/Web technology attitudes
- Externalizing costs by using access opportunities at work

A factor here is also intrinsic attributes, which are an integral part of the physical product and have a direct impact on functional performance (e.g., design features). However, shopping on the Web will likely make it more difficult for a user to evaluate based on intrinsic features, thus extrinsic attributes (e.g., price) should play a more important role (cf. Smith, 1996). This suggests the testing of the following:

> *Research Proposition 1:* Extrinsic attributes will play a higher role for higher priced products than intrinsic attributes; in particular,
>
>   1. price and brand will be the most important attributes used to evaluate a product; and
>   2. recommendations from others (e.g., friends) will be taken into careful consideration.

The previous discussion indicates that extrinsic attributes play a more important role in how people select a product on the Web than intrinsic attributes of the product. However, to the best of my knowledge, how this plays when using the Web has not been investigated.

Externalizing of Costs for Internet Use.   Economists suggest that certain actions by individuals result in positive and negative externalities that affect others (e.g., DeSerpa, 1994). Externalities can also be conceived of as unpriced by-products of (inputs to) a person's behavior (Wijkander, 1985). Negative externalities may require some action by the government to tax certain users (e.g., polluters) whereas in a health insurance scheme, an attempt might be made for reducing free riding on the part of some consumers or clients (Wirl, 1994).

How does this apply to Web usage? We have no empirical evidence to suggest that externalizing of costs leads to different behavior and attitudes on the Web. However, we do have data indicating that telephone pricing does affect the originating of traffic, that is, the product is sensitive to pricing (e.g., Donzé, 1993). An unpriced by-product for an employee with Web access at work is using the Web after working hours for private use. As outlined in chapter 3, by accessing the Web from work after work hours can reduce the employee's costs. First, if the employee has metered telecommunication service from home (fee per call and time), this expense can be saved. Second, if Web access from home to a provider is metered as well (e.g., monthly fee includes 3 hours of access; for each additional minute, Internet access provider charges user), those costs are also likely to be externalized (i.e., saved) by using the employer's facilities to search and download information from the Web. Moreover, if the employee has access to the Web by connecting to his or her account at work from home, Web access charges can be saved again. Finally, in some instances even telecommunication costs are fixed, this is to say that for a fixed price per month, regardless of how often and for how many minutes one calls locally each month (e.g., Canada and United States, see Table 3.1), one can get residential telephone service.

This exemplifies various scenarios on the type of costs one generally incurs for consuming or working with the Web (see also Table 3.1). Moreover, some of these costs can be externalized by using facilities either at work or from home to access one's Internet gateway at work during weekends or evenings. In turn, this suggests that employees may use Web access at work during the workday or during off-work hours for personal business.

Research Proposition 2: Users whose Web access is being metered (telecommunication and Internet access) will likely try to externalize costs by using work facilities as much as possible for nonwork related Web use by

1. using Internet facilities from work;
2. using the facilities less often from home,
3. possibly using call-back features when using workplace Internet facilities from home.

Internet Costs and Web Use.   People will not only try to externalize costs as described earlier, but pricing will also influence the amount of time people spend on the Web. For instance, Coffey and Stipp (1997), using an U.S. sample, reported that about 10% of people watching TV in their study were on the Internet at the same time, based on PC meter data. McDonald (1997) argued that because people use TV and PC sometimes simultaneously, there is little danger for advertisers that TV watching will drop substantially due to more extensive PC use. These authors do, however, ignore the fact that U.S. respondents may behave this way because they enjoy fixed costs for Internet access and telephone charges by the local telco provider. In turn, this permits a user to take advantage of TV and Web entertainment simultaneously (e.g., visiting a TV shows Web site) without incurring additional costs.

Another concern for users has become privacy on the Internet. New technology such as using cookies that store information on the person's hard-disk provide Web site owners with information about a user's preferences, which in turn might affect the advertising or content offered to the individual visiting the site. Moreover, e-mail numbers are being used extensively for direct marketing purposes or for sending unsolicited sales

messages to users (also called spamming). It is likely that users having metered access to the Internet will be more concerned about spamming because it increases their communication (e.g., telecommunication) and
Internet costs (payment to Internet access provider). This warrants further
investigation.

> *Research Proposition 3:* Users whose Web access costs are variable (Internet
> provider and/or telephone/cable charges are metered) will, compared to
> others,
>
> 1. spend less time on the Web;
> 2. have less positive Internet/Web attitudes;
> 3. visit their favorite sites less frequently and during off-peak hours;
> 4. be less likely to purchase or at least to have tested and purchased a
>   product via the Web less often, and
> 5. feel more negatively about direct marketing mail (or spamming)
>   than others.

Spamming and Privacy Concerns.   Recently, spamming types of
messages in one's incoming mailbox have become a nuisance for some people. As outlined in chapter 6, individuals may perceive spamming as a moral
issue and may request protection from such material or they may feel that it
is a conventional issue (see Table 6.1) and thus depends on local customs in
how spamming-type mail should be dealt with. Nevertheless, it seems likely
that to people with greater volumes of e-mail and Web use, they might feel
more negatively about such type of mail. For instance, Gattiker,
Schwenteck, et al. (1997) reported that high e-mail volume users were more
likely to feel annoyed by direct marketingtype messages than others and to
perceive them as a nuisance. This suggests that the following issues be addressed:

> *Research Proposition 3:* Users who perceive their e-mail volume as
> substantial while spending more than average time answering and
> composing e-mail as well as using the Web will, compared to others,
>
> 1. perceive spamming as wrong,
> 2. feel that spammers or direct marketers may have violated their pri-
>   vacy,
> 3. use various means such as filtering software to eliminate spamming
>   mail, and
> 4. ignore requests for information on Web sites or if forced to provide
>   such information in order to receive a particular service, will falsify
>   data.

Other Factors

Although economic factors and information processing may affect Web behavior, other factors may be important as well to explain why users may differ in what they enjoy and do on the Web.

Technology Adoption.   Past research has suggested that some individuals are more likely to adopt technology early, similar to firms, whereas others are more skeptical and might resist new technology for various reasons (e.g., Rogers, 1962). In the context here, I am primarily interested in how individual use of the Internet may in part be explained by being an adopter of other technology. Moreover, Igbarria and Parasuraman (1989) reported that if employees use office technology and perceive it as being helpful, they are more likely to accept the technology without much fuss.

In most countries, mobile or cellular phones are used by less than 20 if not by 10% of the population. Similarly, palmtop computers and digital cameras are still used by a minority of consumers compared to working with a PC or using a camera with film. Such early adopters' attitudes toward and beliefs about the Web should differ from others. Hence, the following research proposition warrants investigating:

> *Research Proposition 4:* It is likely that early adopters of cellular technology, palmtop computers, digital cameras, and information technology (IT) will more likely differ from others as far as the Web and its use is concerned, in that they
>
> 1. have more positive attitudes and beliefs;
> 2. have more experience in testing and/or purchasing a product or service online;
> 3. spend more hours surfing compared to others; and
> 4. are very concerned about privacy issues when using the Web (e.g., cookies, Java applets, providing credit card numbers, etc.).

Sales Channel.   Another concern is, of course, how people might evaluate a product differently based on price level due to the sales channel being used. We were unable to find literature addressing this issue directly. However, literature with consumer goods has addressed the issue about which factors influence purchasing decisions. Jacoby, Szybillo, and Busato-Schach (1977) reported before making a product selection, only limited information was considered by subjects such as brand name and price. This relates to proposition 1, which suggested that extrinsic attributes would be used extensively when shopping on the Internet. However, this difference may also repeat itself when comparing Web shoppers with their counterparts going to a retail store (e.g., Chang & Wildt, 1996).

This suggests that a comparison between shopping on the Web versus in a store might provide answers to the following:

Research Proposition 5: Extrinsic attributes will play a higher role for Web shoppers than for their counterparts purchasing a product at a store; in particular

1. price and brand will be important extrinsic product attributes used to assess a product;
2. recommendations from others (e.g., friends) will be important, and
3. warranty, service (e.g., length of warranty and money back guarantee if not satisfied), and security/privacy issues (e.g., for payment and financing) will be taken into careful consideration.

This suggests that people might have different needs for information when shopping in a store (more traditional way of doing it) versus doing electronic commerce (e-commerce) on the Web. Accordingly, extrinsic and intrinsic attributes may be ranked differently by users when comparing purchasing a product in a store versus via the Web. Unfortunately, research addressing these issues is again lacking.

Private and Corporate Shopper.   As outlined earlier, definite stakeholders could be private and corporate parties who are either current customers or who could potentially become a firm's client (see Fig. 7.1). To illustrate, a person shopping as a consumer uses different criteria for selecting what is important content on the Web site than an individual shopping as an employee for the firm (cf. Rosen & Olshavsky, 1987). Literature has also indicated that purchasing agents are very much concerned about delivery and pricing (Bunn, 1994). Hence, the following research proposition warrants investigating:

Research Proposition 6: It is likely that the following differences between corporate and private customers can be found, such that

1. business users' important Web application will be to obtain information in contrast to private users who will also look for entertainment and freeware;
2. business users' important criteria when shopping on the Web will be quality and brand whereas consumers will be more greatly influenced by price and warranty; and
3. business users will put great importance on delivery conditions (e.g., in stock and delivered within a week).

Although this proposition relates to propositions 1 and 5, it points toward another important distinction to be made as illustrated in Fig. 7.1, namely between private shoppers and corporate purchasing professionals.

Sociodemographic and Cultural Variables

The earlier discussion illustrates how economic factors and information processing by the Web surfer can affect behavioral outcomes. Moreover, other factors such as being an early adopter and shopping at a store instead of on the Web may also influence how people select products with the help of various criteria. In this section, I focus on individual demographics and cultural variables that might again explain differences between subjects' behavior on the Web.

Demographics

Sociodemographic variables such as age, gender, and hierarchical level have been considered important variables in management, psychological, and sociological research (e.g., Bikson & Gutek, 1983; Zedeck & Cascio, 1984). This stream of research has documented results that indicate that demographic variables are often significantly associated with computer and work-related attitudes, as discussed earlier (e.g., Gattiker & Nelligan, 1988). Women's assessment of computer behaviors are generally also more careful and considerate of how it might affect others than men's (e.g., Gattiker & Kelley, 1999).

Some research has also reported differences based on age. For example, older individuals usually are more concerned about moral issues and the welfare of others than are younger people (see Rest, Thoma, Moon, & Getz, 1986 for an extensive review of this issue). In turn, older individuals are also more concerned about computer viruses than are their younger counterparts (Gattiker & Kelley, 1995). There is no empirical evidence to suggest that age differences lead to differences in how people feel about the Web. However, age differences in user attitudes as far as computers and information systems are concerned have been investigated. For example, Igbaria and Parasuraman (1989) reported that older managers' attitudes toward computer technology were more unfavorable and significantly different from younger managers. Kelley, Gattiker, Paulson, and Bathnagar (1994) found a positive relationship between age and respondents' attitudes regarding an information system's ease of interaction (e.g., interactive commands). Extrapolated to a Web context, this research suggests that older users' judgments of Web-based shopping will be different than those of younger users. Thus, the following proposition is offered.

Research Proposition 7: Beliefs and Web behavior are influenced by demographic variables, specifically:
1. Women will differ from men by:
    (a) spending less time on the Web,
    (b)being less likely to have positive beliefs,
    (c) being more concerned about privacy,
    (d) assigning more importance to person-to person contact/communication.
2. Older respondents will differ compared to younger counterparts by:
    (a) spending less time on the Web,
    (b) being less likely to have tested and/or purchased via the Web,
    (c) visiting a Web site less frequently,
    (d) doing more surfing at home than at work.

The Etic Approach.   In a review and synthesis of studies about comparative management, Ronen and Shenkar (1985) reported that part of the variance in employees' work goals, attitudes, and behavior was due to demographic variables, rather than to the individual's country. This would suggest that these measures could be generalized across cultures. Hence, gender influences reported in Canada should be replicated with U.S. data. In turn, this would permit an *etic* approach, which is being accomplished by describing a social phenomenon as relatively culturefree (i.e., using variables that could be generalized across cultures). In this case, gender, demographic and structural influences on Web behavior would be the only variables explaining different behavior. However, some researchers have questioned this and suggest that some differences will remain (e.g., Bhagat & McQuaid, 1982, p. 655; Gattiker, 1992; Griffeth, Hom, DeNisis, & Kirchner, 1985).

Figure 7.2 also includes moderating variables in order to describe Web behavior by respondents across national boundaries. By controlling for the factors outlined earlier (see, for example, proposition 7), it might be easier to explain any cross-national differences obtained. For instance, research by Gattiker, Schwenteck, et al. (1997) revealed that U.S. respondents were most concerned about privacy and security on the Internet compared to German respondents whereas Canadians where in the middle of the two. Thus, the following proposition should be tested.

Research Proposition 8: Cross-national differences will be limited if various socialization, cognitive, and situational factors are being controlled for as far as Web use and e-commerce is concerned; nevertheless, U.S. respondents will put greatest emphasis on

1. privacy and security issues,
2. extensive warranty and "total satisfaction or your money back" offers, and
3. Web use at home during off-work hours.

The role of national cultural differences has long been recognized to have strong affinity on the design of organizational responses (Hofstede, 1989, 1991). Van Maanen and Bealey (1985) suggested that certain types of technological change, such as incremental innovations that simplify existing technologies, may lead to the demise of existing organizational and occupational subcultures. However, more radical technical advances may actually empower old subculture (often suppressed by the overall corporate culture) or may create new ones. Internet technology, it is argued, offers potentials for accomplishing this. Hence, it seems advisable to test the following proposition.

> *Research Proposition 9:* Differences across organizations will be limited if various socialization, cognitive, and situational factors are being controlled for as far as Web use and e-commerce is concerned; nevertheless, respondents from U.S. organizations will differ from other firms by
>
> 1. putting emphasis on privacy and security issues;
> 2. using extensive warranty and delivery services as important criteria for purchasing decisions;
> 3. using Web facilities more extensively at work for employee-related purposes than using it elsewhere.

This proposition is derived from the literature that indicates, as discussed here, that U.S. respondents are usually very much concerned about privacy and security issues in comparison to others (cf. Stone & Stone, 1990; Stone, Gardner, Gueutal, & McClure, 1983). Moreover, marketing literature suggests that U.S. customers are the most demanding as far as service and warranty is concerned (returning products for full refund after having used them for a while because customer is dissatisfied). Moreover, lower telecommunication costs would encourage extensive use of the Web at home in the United States. From a corporate perspective, U.S. organizations are at the leading edge in using the Internet and the Web for corporate purposes.

However, how these issues apply to the Internet and Web context particularly as far as e-commerce is concerned still awaits testing.

SUMMARY AND CONCLUSION

Extensive use of stakeholder theory in research dealing with the Internet and Web issues is quite new. Hence, a systematic investigation of its application to organizational settings is too limited for a theory about its applicability to effective use of Web opportunities (e.g., Freeman, 1984) and its effects to have evolved and received general acceptance. As a result, the propositions set forth were not derived from a generally accepted theory. Instead, they were pieced together from theories grounded in communication, management, sociology, psychology, information systems, and marketing literature as well as from research dealing with Internet issues, extrapolating only when it seemed reasonable.

This chapter also illustrates that based on the issues outlined in the earlier chapters of this book, a framework can be developed to study Internet and Web-related responses of users better. Unfortunately, the information we have at this point about such behavior is extremely limited and is not satisfactory for users and providers of Internet services or for Web sites. Testing of the model and subsequent propositions outlined herein is needed.

A theory may be defined as a set of related propositions that specify relationships among variables (cf. Blalock, 1984, chap. 1). The propositions set forth in this chapter relate to one another (at the very least) through possession of a common independent variable, Web behavior (behavioral and attitudinal outcomes), and thus pass this definitional test of a theory. Yet, more should be expected from a theory, such as a framework that integrates the propositions.

The propositions in this chapter serve as building blocks for the development of a less atomistic, more conceptual theory dealing with Web behavior and its use in business. Moreover, the framework presented must be subjected to review, critique, and discussion across an extended period before gaining general acceptance. However, it is important that the research be carried beyond the traditional discussions and disciplinary boundaries of these issues.

An important conclusion drawn from the research on consumer marketing, psychology, and information systems as well as work sociology is that limited cross-feeding and integration of research results is occurring between work in information systems and computer use, personnel psychology, and consumer decision making as well as communication studies. Unfortunately, such limited exchange of ideas across disciplines will further hamper our understanding of how people use the Internet and the Web in particular. Most importantly, to take advantage of the new opportunities being offered by the Web to organizations as well as individuals requires a better understanding of how people feel, what they believe, and most importantly, how this affects their behavior when using this technology. Researchers should investigate the issues outlined here while using an interdisciplinary approach to better explain various phenomena occurring on the Web (e.g., Blalock, 1984, chap. 6).

One of the most exciting implications for businesses culminating from Web technology is the potential for reviving a more personalized service for customers. This can only be compared to the small local grocery store or textile manufacturer who sold directly to customers. Moreover, the manufacturer knew customers quite well as far as their product expectations and preferences were concerned. In the age of mass distribution and retailing as well as globalization, this has been lost to a large extent, unless the customer is willing to pay the price for such personalized service. To illustrate, on Friday, August 21, 1998, I sent a message to the postmaster at Rollerblade inquiring about the brake on my new inline skate I had purchased the previous week at Big 5 Sporting Goods in Montclair, California. The message was passed on from several people (as I could see on the response) but, most importantly, by Wednesday, August 26, I had an answer. The latter included a to-the-point explanation about my problem with a phone number in Copenhagen and a name of a person I could call and talk to if I needed additional help and a replacement part. I called this individual Thursday morning and he already knew the whole story and was ready to ship the part to my snail mail address as provided in my e-mail signature box. This service was far more convenient for me as a customer than having to pick up the phone and call the United States and being passed on to several individuals. I was happy with this personalized service and my problem was taken care off within a week, even though the manufacturer was several thousand miles away.

The Internet allows firms to revitalize the personalized dimension of retailing, distribution, and servicing of clients by being able to do one-to-one selling irrespective of the customer's location. In turn, this will have some serious implications for the existing production and distribution chain and logistics from the producer to the customer or retailer. This chapter outlines some of these issues that must be addressed to help firms and organizations take better advantage of these opportunities.

8 Final Thoughts and Conclusion

OVERVIEW

As with every other journey, the one we pursued in this book is coming to an end. Here, I summarize some of the many issues remaining open for discussion and requiring our attention today and tomorrow. I start by discussing the opportunity and challenges we face with the Internet. Moreover, how the Internet may affect users such as business, government, and consumers is discussed and the potential success of efforts to create virtual communities based on issues outlined in previous chapters is assessed. Naturally, cultural, economic, political, and regulatory differences, as well as moral, privacy, marketing, and other issues discussed in this book all need further evolutionary development if the Internet is to move toward institutionalization. Finally, implications for public policy, decision makers, Internet users, and researchers are outlined.

This book discusses how ethical issues, communication models, regulation of telecommunication markets, and Internet access, culture, and also cyberspace culture, aesthetics, and marketing all contribute to and draw upon the Internet as a resource or phenomenon. Moreover, as a social reality, the process is continuing by which actions are repeated (e.g., the way we communicate or do marketing using the Web) and given similar meaning (e.g., violating or protecting of privacy) by oneself and other Internet users. The possible development of institutional norms, using the example of privacy, is outlined in this book, suggesting that we have many challenges left before we can assume the Internet has acquired characteristics of an institution such as a multinational firm or the United Nations and its agencies.

If we are using the Internet as a resource for or to broadcast information while also to sell various products and services with the help of this technology, some rules and norms might develop that are adhered to by some (e.g., consumers rights when purchasing on the Internet). Accordingly, it seems unlikely that the Internet will get institutional characteristics beyond national borders except in cases where governments have pushed regulation beyond national borders (e.g., possible harmonization between U.S. and EU laws about privacy or consumer rights on the Internet). Nevertheless, the Internet does offer many opportunities and challenges to users worldwide. This final chapter addresses some of these developments and their implications for users, researchers, decision makers, and policy specialists.

CYBERSPACE: OPPORTUNITIES AND RISKS

Jensen (1990) emphasized that all "new" media were initially seen as a source of utopian benefits and as the precursor of destruction of traditional values and ideals, detracting from church, education, politics, and family. CIS and cyberspace are again challenging some of these values and because of their potential for removing geographical boundaries and spatial barriers between distinct and geographically different places and people, its effect on culture and postmodernism might be substantial.

Table 8.1 lists some of the potential opportunities and risks we might face with the rapid diffusion of CIS and cyberspace penetrating the private and public spheres of our lives. The points in Table 8.1 are not listed in any order of importance; they do not include all the issues that require further discussion. The cultural challenge will be to explore the opportunities to the fullest while limiting the damages. For instance, although today's end-user can consume information, he or she is also able to provide or broadcast information (cf. chap. 3). How we manage the issues in Table 8.1 will also influence the marketing and commercial use of the Internet (see chap. 7). Finally, how we deal with the issues outlined in Table 8.1 may differ based on culture (cf. Fig. 4.1 and Table 4.1), which may result in misunderstandings between users and conflicts among governments. The more structure and diffusion the Internet entails, the more similar the opportunities and risks will be for users around the globe.

Advancing or Decreasing Opportunities.   Table 8.1 suggests that people may have the opportunity to experience new things both visually and cognitively. For instance, an historical site may be viewed with a live cam-

TABLE 8.1

Potential Opportunities and Risks With Cyberspace Culture

People will be Able to Advance/Improve or Increase/Decrease:	People will Face the Challenge/Risk or Opportunity to Cope With a Situation Offering Them an Increase/Decrease of:
experience (visual, cognitive, and others; see chap. 5)	loss of reality, social relatedness, friendship (see chap. 5)
creativity and imagination (see chap. 5)	influence and wealth (see chap. 7)
communication and knowledge transfer through the processing of additional information (see chap. 3)	too much of too little information and productivity (see chap. 3)
independence and loneliness (see chaps. 3 and 6)	Internet or cyberspace use/addiction (see chap. 6)
cooperation and dependence (see chaps. 4 and 5)	privacy (see chap. 6)
efficiency and effectiveness (see chaps. 2 and 4)	variety in consumer demands (see chap. 7)
direct participation in the political process (e.g., through electronic town hall meetings with the president and/or elected representatives; see chaps. 5 and 7)	dependency on having the financial resources to afford surfing the Net (see chaps. 2 and 3)
access to data and information (see chaps. 1, 2, and 3)	confusion and apathy (see chap. 4)

Note. This table benefitted from discussions with Heiner Benking.

era. Hence, pictures about current excavation efforts might be put up on the museum's Web site, thereby offering people from afar a view of the extensive work being done to recover old artifacts from an ancient culture.

The Internet also offers us new opportunities to be creative and imaginary, hopefully resulting in innovative uses of the Internet to improve people's lives. Additionally, communication should improve because with the help of e-mail and the appropriate software, it is becoming easier to reach people in distant places. In turn, this helps people keep a relationship or friendship alive by staying in touch over the months or weeks of separation. Knowledge can be transferred easily by shipping a file to another department electronically.

Communication and knowledge transfer (see left column of Table 8.1) might improve with additional information obtained. However, even if we choose to watch TV news for 3 hours one evening, we are not necessarily understanding and comprehending the issues addressed better than if we had read a high quality newspaper that day for one hour instead. Similarly, the Internet may not provide one with the quality content one desires. Time wasted searching for the appropriate information may ultimately reduce the additional knowledge and insight obtained. In contrast, we might better spend our time by using another medium such as a magazine instead.

Internet users are, naturally, as Table 8.1 suggests, in the position whereby they can increase their independence from others. Instead, a person's dependency (e.g., emotionally) may increase by staying in touch regularly and sharing feelings, emotions, and personal secrets via the Internet with a person far away that one may not see more than once a year.

Cooperation can improve as well with the help of technology by getting people to work virtually while physically being in separate locations. Nonetheless, cooperation might also decrease due to the out-of-sight out-of-mind phenomenon where people forget about another person if they don't see them regularly. An intimate relationship may also suffer even while keeping in touch via the Internet. Hence, the need for physical proximity (e.g., discussing plans during one of the meetings) cannot be replaced with the Internet but, instead, we might be able to reduce the unproductive or somewhat unnecessary meetings by using the Internet to communicate informally and regularly. Naturally, a firm's research and development teams located around the globe may still use travel or video conferencing or other means to have an opportunity to "meet" other team members "personally."

Efficiency and effectiveness can improve with the help of this new technology or it can be reduced. For instance, working requires concentration whereas receiving and answering e-mail could interrupt this process or distract one's train of thought. Similar to the interrupting phone call, people have a tendency to look at new incoming e-mail right away. Instead, by composing and checking e-mail only during certain time blocks in one's working day, we should be more productive than interrupting what we were just doing to check incoming e-mail. Hence, similar to switching one's phone to voicemail, maybe one should switch one's e-mail to be checked once or twice a day. The advantage of e-mail in comparison to the phone is that e-mail eventually ends up being read by the party it was addressed to (or the secretary checking that person's e-mail account). Therefore, the phone tagging experience and waste of time and resources is eliminated.

Although there might be a possibility for electronic town hall meetings, the ruling of the majority might very well end up with the ruling of the mob as one reviewer of this book pointed out. People who are on the Net and can afford time and other resources (e.g., money to log onto the Internet with a Web browser) required for participating in electronic town hall meetings may have more influence than others. Accordingly, a system of direct democracy whereby people participate in the political decision-making process through national referendums might still be the better alternative to having electronic town hall meetings. Nevertheless, electronic town hall meetings do give some groups and individuals a chance to voice their opinion in the hope of making a difference in how their elected representatives may proceed when drafting or passing new legislation or even amending existing laws.

Access to data and information may change as well for many. For instance, people can reach a rapidly growing number of virtual libraries for obtaining information from their home. Naturally, with the help of the Internet and the Web, individuals are better able to find information by using the virtual library while remaining at their desks. Hence, high quality and great depth of content of data and information may be available at one's fingertip from home or on the road by "visiting" a virtual library.

Coping with Situations.   As Table 8.1's right-hand column suggests, surfing on the Internet may also be a problem for some individuals because they lose touch of reality as far as their social environment is concerned (e.g., family, friends, and coworkers). Spending much of one's spare time in cyberspace for playing games and doing other activities, naturally, takes time away from building relationships and spending time with one's immediate social environment. Withdrawing from one's surroundings does, however, also affect one's immediate friends. For instance, we do know from our own lives that watching TV while eating dinner as a family affects communication patterns. Similarly, eating in front of a computer terminal instead of sharing a meal with one's family or friends reduces one's social interaction. One might become too engrossed in the surrealistic world of cyberspace thus losing touch with one's children's worries about school and friends. For instance, children may miss the personal attention and love or not getting one's time required for talking about their experiences at school today because one is too busy surfing on the Internet or watching TV. In contrast, information obtained daily or weekly via an Internet newsfeed might help the individual to better grasp geopolitics and other important matters affecting one's life, thereby seeing matters more realistically. Such a newsfeed may in turn allow the individual to spend less time on the Internet searching and more time doing other things such as raising one's children.

Although people's participation in the political process can improve by providing them with opportunities to communicate with their elected representative, or to share their ideas and beliefs with many by lobbying using the Internet, some may gain a disproportionate amount of influence in comparison to their numbers. We know that certain groups may be the darling of the media and staging effective media events (e.g., hanging a poster on a bridge and chaining oneself to the bridge). Hence the Internet just offers another avenue for influencing decision makers in addition to mass media such as newspapers and television. Moreover, some groups or individuals may lose wealth or gain more with the help of the Internet (e.g., e-commerce). Moreover, people not having access to the vast pool of resources and information on the Internet might be at a significant disadvantage to earn a living.

The difficulty is to find the information one needs quickly or at all on the Internet. Unless a site with the information one needs is listed in a search engine, one is unlikely to find it, except if one has received the link from a friend or has the bookmark saved already. Naturally, links may change and thus bookmarks may no longer work. Search engines may also list sites according to amount of traffic or, in some instances, according to advertising dollars spent and paid to the search engine. Accordingly, one may not come across a short description of a small but interesting Web site quickly. Instead, the site may be listed only after one has browsed over 100 links listed beforehand by the search engine. Hence, some users may end up confused or apathetic and disenchanted after having "surfed" on the Internet for a while, not finding what they wanted to find.

As the previous sections might suggest, a person may, thanks to deregulation of telecommunication markets, obtain relatively inexpensive access to the Internet from home. Naturally, the negative side might be that the individual spends too much time on the Internet. Here the question will be how much is too much? Today's society has become used to people being addicted to many things including television and painkillers besides alcohol and drugs. Again, the individual will have to decide carefully what is enough and what is too much. In the United States, some data has already suggested that people reduce their time watching TV in order to spend more time on the Internet (Coffey & Stipp, 1997).

New technology enables marketers and others to gather much information about a surfer such as the sites visited during the last few days as listed in the system's cache. Moreover, some systems require the user to accept a cookie (e.g., Rocketmail) that in turn provides the other party with information about a person's Web habits, preferences, and other data. Unless the user becomes more aware of these issues and refuses cookies as well as tries to protect his or her privacy, the latter may be lost before one is fully aware of it.

As discussed in chapter 7, the Internet may create new demands by users for certain services. In turn, consumers may demand better services from their local vendors, because the Internet offers them new opportunities to shop worldwide for the best value. On the less positive side, variety for products and services offered may also decrease because the Internet may accelerate to use of global brands and services. For instance, TV fare such as documentaries by a local channel might be replaced with a National Geographic specialty channel broadcasting documentaries 24 hours a day. Such information provided may, naturally, represent a cultural bias. Accordingly, a U.S. TV crew filming events or English journalists putting together an Internet newsfeed will interpret events according to their own cultural biases (cf. Fig. 4.1) and may in turn provide viewers and readers with different content than their French or Chinese counterparts.

As Table 8.1 also suggests (right column), people may depend to a great extent on their financial resources that in turn may determine if they are able to spend much time at the Internet. Moreover, being employed may ease this burden because Internet access from work or by connecting to one's account at work during off hours reduces the financial burden put on some user groups for taking advantage of the Internet.

Communication and knowledge might improve with the help of the Internet and the Web, however, there is no guarantee about the quality of some of the content. Moreover, for many of us, we might simply experience too much of too little. Sitting in front of a computer while being connected to the Internet may neither improve our social skills for interacting and functioning in a group, nor make social life more exciting. We may be able to be in contact via e-mail or discussion groups with people from near and far. Unfortunately, interaction via the Internet is neither as rich nor as full of surprises and nuances (e.g., nonverbal communication, facial expressions, and eye contact) as is a telephone conversation. Most definitely, a personal meeting between two individuals who chose to be in the same geographical space at the same time is the richest form of interaction if people have the skill required for communicating in person with another human being.

IMPLICATIONS, OR, WHERE
DO WE GO FROM HERE?

The previous section illustrates that the Internet offers us opportunities as well as challenges and risks to be better or worse off than without this technology. But regardless of the opportunities and risks, Internet use is spreading. Some have compared the rapid diffusion of Internet technology with the TV. The U.S. data indicate that the use of the Internet by households has outpaced the growth of TV use in a 5-year period. For instance, in 1993 (1948), less than 250,000 households (just about 250,000) had Internet access (a TV). In 1997 (1952), it was estimated that 21,000,000 (17,000,000) households had Internet access (a TV). Although TV was available before 1948 (1939 World's Fair), U.S. factories where not really able to switch their radar tube to TV tube production until the war ended in 1945. Also, in 1952, unlike today, a TV set was virtually equal to one household ("Commerce Department's TV vs. Net", April 29, 1998).

Naturally, although growth of the Internet in private households, thanks to graphical Web browsers, has outpaced the growth of TV sets in the late 1940s, our interest has been more on the potential impacts the technology might have on our lives at work and elsewhere. This book suggests that with the help of telecommunication, satellite technology, and much more, users have more opportunities than ever to watch various TV channels and programs. As chapter 6 shows, however, the variety of the fare offered has not necessarily increased but the influence of American TV entertainment has reached the last corner on this earth.

In contrast, the Internet offers the user a variety to choose different content but, most importantly, to create his or her own content to be shown on a web page. Alternatively, a person's information can be distributed via an electronic newsletter to as many interested parties that are willing to sign up and, most importantly, regardless of their location. I have little choice about which TV channels are fed into my house through my local cable connection except maybe choosing between two or three standardized packages. Even with a satellite dish, I can watch the channels that are being beamed over my neighborhood only or for which I can purchase and afford a decoder. With the Internet, the variety of options for entertainment and information as well as the research content is much greater than with satellite TV and, most importantly, access can be relatively inexpensive (see chap. 2). Nonetheless, as the following sections show, some challenges still remain.

Regulation, Policy, and Access

In this book, I state that policy, regulation, and access to the Internet are important. Nonetheless, we must understand these issues in conjunction with cultural, ethical, and moral similarities and differences between countries. This might help us in better understanding why certain e-commerce and marketing applications might work in some settings but flop in others. In the context here, we are interested in discussing some of the implications for users, organizations, and governments.

User. For users, a well-developed and generally adhered to regulatory framework for electronic commerce would be advantageous in several ways. For starters, it would make it easier for users to choose payment and transaction methods that are generally accepted and considered safe. Accordingly, using an electronic wallet or making micropayments via the Internet (e.g., less than $1.00) would no longer depend on the compatibility of systems, legal concerns, and other matters. Another concern for users is their privacy. Here, privacy standards must be adhered to and abided by government agencies, business, and private users. Unfortunately, security agencies often side-step rules and regulations violating privacy without the affected parties being necessarily aware of it (e.g., by using Tempest as explained in Appendix A). Without having a court order, a person's privacy may be violated. The individual may have no idea for what, by whom, and when such illegally obtained information may be used. Hence, if a user is not aware of when and how his privacy has been compromised illegally (e.g., files and data), it is very difficult to protect one's right to privacy.

One of the best defenses for Internet users is to assure that one is keeping oneself informed and abreast new technical and regulatory developments. One way to do this relatively easily is by signing up with a moderated electronic newsletter addressing these issues. Moreover, normal precautionary measures would also be a sensible strategy, in turn limiting the potential for misuse of one's information about shopping on the Internet. Hence, the individual should be careful in providing information about oneself on the Internet or elsewhere without having checked the firm's privacy policy beforehand.

Another concern is that whatever is being sent via the Internet is probably documented for years to come. As much as the White House has been forced to archive e-mail messages sent by citizens, firms are also required to keep backup copies and archive such data in electronically readable form. Moreover, messages may end up with the wrong individual (e.g., postmaster) instead of the addressed party. Hence, using encryption might be a useful way to protect one's privacy as well as a message's content from the wrong eyes.

Organizations.   For this group, the most important implication might be that with the Internet, far away markets may just be next door, while competition in one's backyard will increase as well. Because of this, organizations must undertake the necessary steps to assure that they can remain competitive in this changing business environment.

For some organizations, not all workers are provided with full access to the Internet. For instance, some may not have Internet access at all. Others may be unable to use a graphical browser to take advantage of the Web, despite being given the necessary system privileges to receive and send e-mail. What is appropriate cannot be decided here. Nevertheless, it seems clear that skills about Internet technologies and their effective use will become more and more an issue when hiring workers. Hence, lacking these skills may prevent an employee from being hired or may reduce career opportunities in the firm (e.g., transfer). Moreover, organizations want to make sure that employees keep abreast of new Internet developments while the technology is being used to the firm's advantage. In turn, it seems strategically important to assure that all employees are knowledgeable enough about the Internet and keep abreast of new developments. For this reason alone, the only wise route to take is to provide every worker with access to the Internet. This should benefit the firm as well as the employee.

Although giving every worker Internet access may be strategically wise, some steps must be undertaken to ensure that Internet resources are used effectively. For starters, firms should have a clear privacy and safety policy (see Appendices E and F). In such a policy, it must be made obvious to the worker what one can and cannot do and, most importantly, what consequences one may face if they violate the policy. Moreover, to assure that clients and potential customers trust the firm's privacy and security policies, these need to be published on the firm's Web site and for possible violations, an ombudsman is needed who receives complaints about policy violations from employees or clients.

The firm should also adhere to the policy that unless a person states specifically that his or her information can be used for marketing purposes, the assumption is that one has opted out. Accordingly, no data should be used in any way to subsequently contact the client or to inform him or her about new products. Similarly, none of the customer data can be sold to others and, most importantly, using technical means such as placing a cookie on one's machine would not fit the firm's privacy policy. For clients, it is important to be able to understand the firm's privacy and security policies. In turn, clients or visitors to the firm's web page must be able to trust the organization and, most importantly, the latter must be able and willing to follow these policies strictly while respecting clients' wishes as far as privacy matters are concerned.

On another front, it should be clear to employees that the Internet can be used for private purposes during lunch or after working hours such as weekends. In fact, the firm may want to encourage such use and offer it as a benefit to family members such as children (e.g., additional private e-mail accounts and Internet access from home), because costs may be minimal in comparison to the goodwill created. Nevertheless, private use of Internet facilities at one's workplace during working hours is not acceptable and consequences for doing so must be stated. Most importantly, when an employee is caught using the Web to check out one's favorite holiday spot during working hours, the consequences of such behavior should be enforced as outlined in the policy document.

Unfortunately, during a recent study we discovered that most organizations to not have an explicit policy whereby employees are permitted to use the Internet for private purposes outside of work. Nevertheless, most organizations know that their employees are using the Internet for private purposes and this is usually just tolerated. When working for the U.S. federal government, however, employees are required to avoid using the Internet for private purposes whatsoever. In practicality, however, although not tolerated officially, private use is simply overlooked or ignored unless it gets out of hand (e.g., downloading pedophile files).

All the earlier approaches described do not appear as effective as one would hope. Neither prohibiting private use nor tolerating private use quietly nor restricting Web access make matters easier for the firm or the employees. For instance, disallowing the use of the Internet categorically seems silly because it is nearly impossible to enforce such a rule, thereby resulting in management tolerating or simply ignoring violations by many. Having a regulation on the books that is being broken at least once a day by many employees does not seem not very helpful. Additionally, not permitting or disabling access to the Web will result in employees not acquiring the skills necessary for using the technology effectively, whereas line charges will hardly increase if private use during off-work hours is permitted, if not encouraged (e.g., after 4 p.m.).

Regardless of what a firm tries to accomplish, all employees must be entrusted to act responsibly and encouraged not to abuse or misuse their

Internet privileges. This philosophy should be engrained in the firm's culture and is probably much more effective than having 10,000 rules. The latter force the firm to act in a policelike manner whereas having the appropriate organizational culture makes employees' behavior more likely to conform to policy. Encouraging employees to follow the firm's Internet culture, thereby acting responsibly while making their Internet choices wisely seems a more positive and probably a more effective approach than playing police. Nevertheless, similar to the monthly audit of petty cash, some random checks about employees' Internet behaviors during working hours might be useful.

Government.   As chapters 2 and 3 suggest, we have come far in North America and Europe as well as Japan, Australia, and New Zealand to mention a few places, as far as deregulation of telecommunication markets is concerned. Nevertheless, most cable, power, and telephone operators are somehow linked with each other either through share swapping or by having purchased a company in the other business. Hence, the Swedish telecommunication firm Telia is not necessarily interested in offering very cheap Internet access through its cable subsidiary Stofa in Denmark. This changed once TeleDanmark offered ADSL services at competitive prices, in turn, providing users a cheap alternative to cable Internet access at satisfactory speed for downloading content.

Until telecommunication infrastructure is owned by several firms, competition is limited because all firms may have to lease line capacity from one provider. Except for cellular communication, all telecommunication infrastructures (e.g., lines) including the Internet backbone are still owned by TeleDanmark. To assure fair access to TeleDanmark's infrastructure for competitors, the Danish government regulates the prices charged for renting or leasing lines. In turn, this illustrates that competition cannot really survive without governmental interference. The Danish example applies to many EU countries except for England, Finland, and Sweden, where competition is already in full swing, with different firms owning different parts of the infrastructure as well.

Another development can occur whereby several firms specialize in building, maintaining, and owning infrastructure whereas specialized firms may sell telecommunication services only. This development is a new trend in the United States for electric power. For instance, a city or a firm may purchase power from a wholesaler who in turn rents distribution infrastructure to bring the power where it is needed. In turn, prices are likely to fall by offering clients to purchase from the local utility company or a wholesaler depending on prices. Similar developments are happening in the cellular phone business whereby a firm such as Debitel (member of the Daimler Benz Group) has become the biggest reseller of these services in Germany and Denmark although, most importantly, it doesn't own its own physical network. Debitel's rates are highly competitive and have put pressure on the owners of the cellular networks to lower their prices for cellular services to avoid losing business.

Accordingly, telecommunication and cable as well as satellite communication services may become more similar to other industries, whereby the manufacturer or producer may sometimes sell to the user but specialized sale companies may spring up with brand names dealing with the final user. Only the future will tell if this will reduce prices further. Experience with the cellular phone business in Germany and Denmark shows that wholesalers are reselling cellular capacities at very competitive prices. In the United States, wholesale firms purchasing electric power through auctions and selling it to various user groups has also reduced energy prices. Hence, if governments allow the development of such markets, telecommunication or communication prices for Internet access in general will, most definitely, come down further.

INTERNET COMMUNITY AND SHARED CULTURE

The previous sections of this chapter point out the many challenges and opportunities the Internet offers to users around the world. Additionally, various user groups including organizations and governments have undertaken some steps to make the Internet work smoothly for all parties even though their interest might be very diverse if not in conflict with each other. Recently, many have used community as a new buzzword for the e-commerce community (see also chap. 4). However, as the following discussions illustrate, their success both commercially and otherwise has not always been forthcoming as easily as initiators might have hoped for.

Virtual Communities and the Internet

Communities are imagined and can be distinguished based on the style in which they are imagined. Communities are generated by communal experiences and as such, culture and language may be an important ingredient for understanding each other. As discussed in chapter 4, however, users on the Internet may communicate with the help of the lingua franca English but because the latter may not be their mother tongue, many are unable to use the language to its full richness. Accordingly, semantics and vocabulary used may be more narrow or simplistic than might be the case in the usual discourse between two native speakers. The latter may also use humor and shared experiences (e.g., during youth, schooling, or cartoons watched as children) to further illustrate their meaning or to portray certain images to the listener. Hence, a community member may share many experiences and exposure to literature and mass media comparable to other members. In contrast, an outsider may lack these and have a hard time understanding some humor passed on via e-mail or references made to a favorite character in a local TV series.

As a mass media, television communicates in one direction to its viewers and many share the same experiences, stories, and laughs if they watch an episode of a famous sitcom or a football match being shown live. In contrast, the Internet permits users to receive information while also creating content and information that in turn can be shared with many. Through hyperlinks, a reader may not even finish reading a document or playing a virtual maze he or she entered but may instead connect to another site. In turn, even people who visit a Web site do not necessarily share the same experiences by proceeding and reading the same content as the next person might. Hence, the world of official power or influence as used by dictatorial regimes by controlling the mass media may be challenged, because users can influence the setting of the agenda for discourse and may thus influence the final outcome significantly.

In a community, the number of free riders is kept as low as possible (see also chap. 5) and certain systems are used to make sure that most members comply to local tax laws (cf. de Jasay, 1989). Problems arise when we are dealing with a nonexclusive good such as health care or unlimited Internet access, whereby users can increase their share of using resources without increasing their contribution to the system. Virtual communities are primarily built on voluntarism, that is, whereas a member may contribute as much as they like, another may consume and thus benefit without contributing much if anything. Free riders are possible as the nonexclusivity of goods further encourages such behavior. This might explain why hacker communities are based on tid for tad, that is, if I give you a package of code or images, what would you offer me in return? Only based on this bartering system will the transaction go forward and otherwise, no deal may ever be consummated.

This illustrates that there is more to a community than virtuality. A community may encompass, but not necessarily be limited to these five components:

1. personal relationships making up a social network;
2. simple and open access to the community for interested parties;
3. personal meetings and understanding of each other;
4. dialogue and feedback as well as shared experiences;
5. a common history.

These components are all characteristic of a community and Internet communities rarely encompass all of these. Nonetheless, people with similar interests and in limited geographical areas may form cyberspace communities about a topic of shared interests (e.g., baking or starting a new business). If the cyberspace community is limited to an area such as Denmark or Northern California, having members getting to know each other better by attending parties, weddings, business openings, and other celebrations might raise community spirit. Although such activities are organized via cyberspace, they happen by getting people together at the same time and geographical location. Naturally, such a community is far different in its objectives, activities, and membership than what some e-commerce consultants might sell to an organization trying to harness its Internet activities in the right direction.

Developing Communities on the Internet: Structure and Community

To illustrate the previous discussion, the Well in Northern California allowed its users to develop a cyberspace community; many of them shared experiences such as weddings, parties, and get-together meetings, whereby participants were able to develop their acquaintances further and to possibly become friends. In the case of virtual communities, meeting in person is often impossible due to geographical distances and just chatting with each other using pseudonyms does not necessarily help build a community spirit.

Local-Area-Network (LAN) parties are organized more and more frequently, whereby participants' computers are linked with each other within a room or building allowing people to play games for many hours. For instance, BeatDown in Costa Mesa (Southern California) offers such a game community whereby instead of permitting people to play over vast distances, people travel vast distances with their computers to enable them to participate in a face-to-face computer game. Hence, BeatDown illustrates that even hard-core gamers require physical interaction to make their virtual community flourish. Such computer parties go on for hours and although nobody sleeps, participants may meet in an Alberta airplane hangar, dance, sing, and, most importantly, play head-to-head computer games for hours.

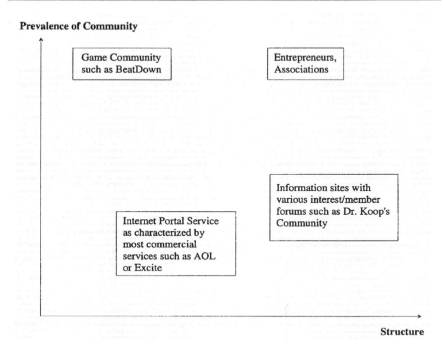

FIG. 8.1. Prevalence of community characteristics and level of structure of an "Internet community."

The previous examples illustrate that for developing a community encompassing all or at least some of the five components characterizing this construct as outlined earlier, face-to-face communication still seems to play an important role (Gattiker & Hedehus, 1999). Figure 8.1 illustrates this further.

The prevalence of community characteristics seems lowest in a situation of a commercial service such as AOL. Although many people may use the network, its members rarely if ever have personal meetings and relationships or a common history. On an information site such as Dr. Koop's Community (the former U.S. Surgeon General), participants may share a somewhat similar history such as being cancer victims or having a member of their immediate family suffering from a cancertype of disease. Such a community offers more structure than AOL will ever offer, thereby permitting "members" to find the information and advice they need quickly.

If the structure is high and most of the five characteristics of a community are prevalent, these individuals share a common interest (e.g., profession, field of work, or hobbies—a community of entrepreneurs or an association's web site). Access to the community is relatively easy and personal relationships may have resulted in forming the Internet community in the first place or else, get-togethers may have resulted in building personal relationships. Dialogue and feedback occur quite frequently while discussing various issues using chat forums or listservers (moderated or nonmoderated). The community has its rules and norms that it expects its members to follow. Quite likely, the community might even have a charter outlining its philosophy, again depicting a high structure. If money for dues or services changes hands, its not-for-profit status similar to a church may have been secured from the tax authorities.

A community not having a definite or rigid structure but, nevertheless, exhibiting a high community spirit might have as its members people who love playing computer games (e.g., BeatDown, see Fig. 8.1). Hence, where people might meet regularly to party or may meet each other not only in cyberspace but, most importantly face-to-face, membership might be loose. Accordingly, people may join or leave a community or as illustrated by BeatDown, membership may change every weekend because people attending are selected out of a pool of many interested players. Consequently, the structure of such a community is low but social networks may be built based on shared weekends of playing games and dancing while not having time to sleep.

SUMMARY AND CONCLUSION

Unfortunately, large traffic sites with extensive financial and other resources (technical and know-how) will be the ones coming up on search engines first. In turn, the information fare one is getting on the Internet might not be as diverse as we might wish. Dominant information providers with their big Web sites may provide standard fare. Already today much of the information produced is created on behalf of business and governments (e.g., press releases, reports, advertising material, etc.). Unless nonpartisan parties such as individuals, associations, and others succeed in making their news and information widely distributed for people, information sources and interpretation of events will not necessarily become more reliable and accurate but simply more unified.

As Table 8.1 would suggest, with every opportunity for positive outcomes, we also have a chance to have negative results instead. It is not so much the technology as the way we use the technology that will shape our information future. A community creating and managing a site or a virtual space will succeed or receive support and/or public or private funds only if there is a demand for its services and content by the larger community out there. As much as a museum or a library has to show that its facilities are used extensively by the public, so does a commercial or a community group's Web site have to show that it serves its clients well. Many visits and extensive downloads of materials stored on the site are two important indicators of how well the Web site is serving a need for information out there.

The Internet has certain similarities to a house and to an organization insofar as it consists of people and structures to organize processes and tasks. An organization's design is "… along various lines of people among social positions that influence the role relations among these people" (Blau, 1974, p. 12). In the context of the Internet, various groups of people and organizations with various positions and needs as well as roles affect the relationship between these various parties.

From a cultural perspective, objective and subjective dimensions of culture (see Fig. 5.1) will also influence the building of cities and houses. For instance, a city built for the car requires large garages and driveways for houses. Also, houses in the Mediterranean are built differently compared to a home in the Canadian Arctic Circle. Moreover, whereas most U.S. houses have a TV room in which one might also find the family's PC, no such dedicated room is usually found in Denmark or Germany. Instead, the PC is in the parents' bedroom or a room being dedicated as the parents' home office. Considering these differences in people's environments, the question must be posed if it is possible to develop a community or at least a community spirit for Internet users in vastly different locations? In turn, we need to discuss whether there is a community spirit as exhibited in virtual communities on the Internet, the first step for having institutional characteristics evolve from this.

People in organizations or institutions may not necessarily agree on what the organization or institution should be like. With buildings, once they are built, they remain the same until a major remodeling may be done a few years down the road (e.g., adding nice washrooms to hotel rooms in a building from medieval times). Accordingly, houses and office buildings today look different than the ones built in the 1960s or 1930s. With organizations, their structures may change due to fads or fashion. Also, organizations evolve and change over time depending on the environment (see contingency theory) and the people working in the organizations.

Appendices

The following appendices provide the interested reader with some additional information, insights, and tidbits about the issues discussed in chapters 1–8 of this book. Most of the material provided should further facilitate the transfer of the material presented herein, and thereby assist in its use and application by readers.

APPENDIX G

Interesting Web Sites for Tips, Hints, and Information

APPENDIX H

Additional Sources for Information

Appendix A

The Internet: Glossary of Terms

Algorithms: All modern algorithms use a key to control encryption and decryption.

> *Symmetric* (or *Secret key*): Uses the same key for encryption and decryption, or the latter key is easily derived from the encryption key. (See Algorithms).

> *Asymmetric* (or *Public key*): A different key is used for encrypting and decrypting a message; accordingly, the decrypting key cannot be derived from the encrypting key. (See Algorithms).

ASCII: ASCII is a universal computer code for English letters and characters. Computers store all information as binary numbers, regardless of what make or brand the computer is. ASCII also refers to a protocol for copying files from one computer to another over a network, in which neither computer checks for any errors that might have been caused by static or other problems. The difficulty with ASCII is twofold: All special fonts and elements used in a document typed using one software (e.g., WordPerfect) will be lost when the file/document is saved in ASCII format and then reloaded by somebody else into WordPerfect (or even Microsoft Word for that matter) and, special characters such as those used in German cannot be transferred by ASCII (e.g., ü is transferred as a blank and a ue must be typed instead). In addition to German, for many languages (e.g., Chinese and Japanese), this represents a real problem and may, in part, continue to limit the use of the Internet in these countries until ASCII is replaced by another standard, easier or more feasible for use in languages besides English.

Asset Value: This term describes the value of an asset such as an information system and/or a database the firm currently has. Investments undertaken using technical and nontechnical measures (e.g., firewall and better use by employees of accepted security measures available) can be compared to the costs incurred if the asset is damaged or possibly even lost.

Asymmetrical Digital Subscriber Line (ADSL): Allows voice, video, and data to be transmitted over a single telephone line at up to 6.144 megabits per second (Mbps) in a single direction, with significantly slower speeds in the other direction. Accordingly, it is very appropriate for downloading large amounts of data but for certain applications, such as videoconferencing, ISDN works better.

Availability of Data: This refers to whether the data and information, as well as the necessary systems, are all accessible and useable on a timely basis as required to perform various tasks. For instance, medical personnel must get access to patient files even during a massive power outage where generators may have to be used to guarantee availability of data.

Browser: An access tool to the WWW that uses hyperlinks to access remote information. Browsing is using the browser to look at the information on the WWW. Cruising and surfing are synonymous with browsing.

Bulletin Board System(s) (BBS): A type of online computer service that functions as an electronic notice board. Users can read or post messages, download programs, and play online games. Some functions of a BBS are similar to that of the Internet, but on a smaller scale.

Cable: Cable is 200 times as fast as a normal telephone line, about 70 times as fast as ISDN, and about 7 times faster than ADSL, while offering the same speed in both directions. Hence, cable companies around the world are starting to offer their subscribers Internet access (if permitted by law) in addition to television channels.

Call Answer Service: This service will actually take several messages at once—even when your line is busy! It answers in your own voice and lets you access your messages from almost any touch-tone phone around the world. Monthly charges apply for service.

Call Forwarding Service: Allows the user to enter a phone number where one can be reached if one leaves one's home. Accordingly, calls will be rerouted automatically to the other phone. Rerouting toll charges are paid by the person registered under the phone number initiating the forwarding of calls, unless it's local and the user pays a fixed monthly fee regardless of number of calls and time spent on the phone (e.g., most areas in the United States and Canada). Monthly charges apply for service.

Call/Name Display Service or *Caller-Number Identification* (CNID), also called *Automatic Number Identification:* You can see the name and number of the person who's calling you by glancing at a small display panel on your phone. It's a convenient way of knowing who the call is for. Monthly charges apply for service.

Call Waiting Service: You won't have to worry about missing that other important call you're expecting, even when your line is busy. A soft beep lets you know when someone is trying to get through. Monthly charges apply for service.

Communities are imagined and can be distinguished by the style in which they are imagined. In turn, a society can be seen as specific instances or styles of imagining community that are founded on communal experiences (Kjaerulff, 1998). In turn, culture and language are important in defining limits for interaction. However, with the Internet and knowledge of lingua franca such as English, the importance of a national language may decline (see also Internet Communities).

Computer Ergonomics: Applies to such things as workstation design, readability of screen and movement of keyboard or tilting of screen to adjust to the human's motor needs (e.g., avoid eye fatigue; see also Ergonomics).

Computer Mouse: A "mouse" is a pointing device to select items on a video screen. It is a small box set atop wheels or ball bearings and attached to the keyboard with a wire to improve maneuverability. Rolling the mouse on the table will cause analogous movements of the cursor on the video display. Buttons protruding from the top of the mouse can be used to select certain commands and actions.

Conference Calling: A registered user can initiate a conference call between two or more parties from one's phone; hence, family members or business partners in diverse locations can make an appointment by being online at one time, avoiding phone tagging. Toll charges, if any apply, are paid by the initiating caller. Monthly charges apply for service.

Confidentiality of Data: This refers to data or information that is not made available or disclosed to unauthorized parties (e.g., individuals, organizations, and processes). For instance, medical data is not being disclosed to others, such as employers, making the identification of the patient impossible unless one is authorized.

Cracker(s): Forms of predatory and malicious behavior, including malicious computer intrusion, are embodied not by hackers but by "crackers" instead. Accordingly, they try to take a peek at data or information and use other hardware and software without legal rights to do so. Unlike hackers, however, crackers cause damage or losses requiring restitution to the property owner (e.g., by damaging of files, illegal copying and distribution of proprietary information and other losses; cf. cyberpunks and hackers).

Cryptography: The science of keeping a message secret.

Cybercash: Represents money digitally and requires an electronic system with virtual credit used worldwide. Some firms are offering cybercash to shop at their "cybermall" using the string of digits (credits) being transferred from the consumer to the seller. Security and acceptability concerns are still many but standards are being developed. How this may affect local monetary policy is still unknown. This is a real challenge because such money floats around and is neither limited to one country nor under the control of any particular central bank per se. Cybercash is not really cash but instead represents an obligation that the issuer has incurred to pay a monetary amount at some future date. In contrast to cash, which a merchant must accept for payment by law, using a stored-value card for settling one's bill can be rejected by the store owner, according to a report by the American Bar Association (Iwata, March 10, 1997).

Cyberpunks: This term embodies behaviors similar to what we define as crackers and phone phreaks but, in addition, they use technology to damage, destroy, or capitalize on the data they find (e.g., for profit) and, most importantly, they use their know-how to gain more information, which, in turn, increases their influence, power, and potential threat on others' information world. Most of this activity is done at somebody else's expense, for example, by breaking and entering into an organization's voice mail system, to log-on long distance into another information system illegally in order to copy information (e.g., medical and financial data), while trying to profit from the information (e.g., selling it to another user/business/government willing to pay). Similar to the things we learned in Economics 101, demand and supply of goods are related. Accordingly, cyberpunks can be compared to "fences" in New York who sell stolen merchandise at very "attractive" prices. As long as there is a consumer who is willing to go home with a "good" deal, even if it is at the expense of the victim whose apartment/car was broken into, then the supply will continue. If we stop purchasing from fences, supply will drop; if cyberpunks find no demand for their illegally acquired information and, therefore fail to sell their goods, supply will be reduced (see also Hackers and Crackers).

Cyberusers: These epitomize individuals who use computers, phones, and cyberspace to derive jobs and exceed current limitations of technology and reality without performing implicit or explicit violations of current laws and ethical as well as moral standards held by society. The interesting thing is that the information highway offers cyberusers to both consume (e.g., sign on at a commercial information provider such as CompuServe) and broadcast/produce information on today's information highway (e.g., edit and distribute electronic newsletter, join online discussion groups, administer a list server of electronic addresses).

Damages: It is difficult to define these in the context of the Internet and bytes and data. However, the Swiss made a first attempt in coming up with a legal tool that should help in protecting one's data and information. Swiss legal infrastructure about data safety/security and computer viruses changed when article 144bis—Damaging of data became part of the country's criminal code; it came into force January 1995 and states:

"1. Anyone, who without authorization deletes, modifies or renders useless electronically or similarly saved or transmitted data, will, if a complaint is filed, be punished with imprisonment for a term of up to 3 years or a fine of up to 40,000 Swiss Francs.

If the person charged has caused considerable damage, the imprisonment will be for a term of up to 5 years. The crime will be prosecuted ex officio.

2. Anyone who creates, imports, distributes, promotes, offers, or circulates in any way programs that he/she knows or has to presume to be used for purposes according to item 1 listed above, or gives instructions to create such programs, will be punished with the imprisonment for a term of up to 3 years or a fine of up to 40,000 Swiss Francs.

If the person charged acted for gain, the imprisonment will be for a term of up to 5 years" (English translation by Figerio, who was instrumental in the legislative process, presented at EICAR 1995, Rüschlikon/Zurich, p. 2; see also footnote 1 in chapter 7)."

Although the article provides quite a comprehensive legal framework for damages, to quantify these in monetary terms is often difficult (e.g., hours of work lost, potential lost sales, etc.).

Decryption: Denotes the process of retrieving the plain text from the ciphertext.

Definite Stakeholders: They have power and legitimacy; additionally, their claim is urgent, such as a government health agency requiring information in case of an epidemic from PMI. Its demands may be given priority by management; for instance, key suppliers, such as doctors, demanding immediate attention (e.g., need for complete medical history on an insured patient to help fight deadly virus discovered in a patient's blood).

Digital Signatures: Public key algorithms can be used to generate a digital signature, which is a block of data used to create some authentication.

>   *Public key:* The public key is used to verify that the signature was really generated using the corresponding private key. Public keys are often registered with a third party and can be downloaded so the person can check if the key is genuine (see Digital Signatures).

>   *Private key:* The party initiating the sending of the document with the signature generated this key; it is needed to generate the digital signature (see Digital Signatures).

Emoticons: These are a clever combination of keyboard characters to punctuate a message with just the right spirit. To read them, one has to tilt one's head slightly to the left (see Table 7.4 for a comprehensive list of emoticons).

Encryption: This is the encoding of message contents thereby hiding the data or information from outsiders; in turn, the encrypted message is called ciphertext.

Ergonomics: This term refers to the science concerned with the study of the functional relationships between human beings and technology such as computers. Ergonomics considers the characteristics of people when designing and arranging technology and work space, thereby helping to increase the effectiveness and safety of interaction. The concern is primarily with physical and sensorimotor aspects of humans and not intellectual aspects. Ergonomics is one facet of the person–computer interface (see also Computer Ergonomics).

Expectant Stakeholders: Three attributes, power, legitimacy and/or urgency are present, have an active stance, and thus PMI is more responsive to an expectant stakeholder than to a latent one.

>   *Dominant Stakeholders:* They hold legitimate claims and power to act on them, such as investors, employees, key regulators, and suppliers.

This group may demand special procedures; for instance, regulators outlining how a citizen can get access to one's records (see Expectant Stakeholders).

Dependent Stakeholders: They hold urgent legitimate claims but need an ally such as small shareholders and/or the public demanding some action in order to be heard. An example is a consumer advocacy group demanding better privacy policies for the health insurer collecting and using patient information (see Expectant Stakeholders).

Dangerous Stakeholders: They have urgency and power but no legitimate claims, such as the Unabomber (or terrorists) being charged in 1996 in the United States; for instance, hackers and spammers who may use various means to unlawfully gain access to medical information records subsequently being used for their own needs and/or requirements (see Expectant Stakeholders).

Extranet: An Extranet helps the organization to link the outside world such as suppliers and customers with a private Intranet. Although this is similar to the Internet, access is controlled and restricted to particular groups, similar to the Intranet. Accordingly, an Extranet Web server can be accessed by all the participants involved in a project (e.g., various engineers and the firm developing a new product), but not by anyone else. In this example, the Extranet provides project management functions for the work in progress and the work teams involved. The security is increased; however, breaches can still occur (e.g., cyberpunk getting hold of a password or access to a server who, with a password, is given access to the Extranet; see also chapter 4 and explanations in this appendix for Internet and Intranet).

Flamed: A virulent and often largely personal attack against the author of a posting on the Internet. Flaming occurs more frequently than is probably desirable.

Freenet: Information network (often a community network) capable of providing individuals with "free" (subsidized) access to the Internet. Some Freenets offer bulletin board and databases for community use and e-mail but without access to WWW pages and voice. Others provide full-fledged Internet access (see also SaveNet below and Table 3.3).

FTP: File-transfer protocol—access to hundreds of file libraries (everything from computer software to historical documents to song lyrics). You will be able to transfer these files from the Internet to your own computer.

Gateway: A connection between two networks. It is necessary for all information to go through a gateway on its way to another network.

Gopher: A menu-driven program that gives access to hundreds of other databases and services on the Internet. They can also be used to copy gopher text files and programs to a computer.

"Hacker": This is somebody who derives joy from discovering ways to exceed current limitations. Thus, hackers in the original sense were referred to as explorers who solved problems and exceeded conventional limits through trial and error in situations where there were no formal guidelines or previous models from which to draw. A hacker tries to take a peek at data, information, and to use other hardware and software without legal rights to do so, thereby exhibiting behavior that could be labeled *trespassing* (see also Cyberpunks and Crackers, and Table 4.1).

Home Page: A document on the WWW that provides information about an individual or organization. It can be accessed through a *URL* (uniform resource locator, which is the address of the page indicating the WWW server on which the page is located). The page can have links to other pages on the same computer or to other computers (so-called hyperlinks).

Icons: A small graphic image representing a link to a graphic, another page, or a program.

Institutionalization: This is the process by which actions are repeated and given similar meaning by self and others. In this process, individuals create definitions of social reality by inventing distinctions or typifications, thereafter treating these productions as objective and external to their own actions. Berger and Luckmann (1967) argued that social reality is a human construction, being created in social interaction (see also Community and Internet Community).

Integrity of Data: This term refers to whether the data and information have been modified or altered in an unauthorized manner. For instance, unauthorized personnel are unable to alter medical records whereas changes made by others (authorized personnel) are tracked and recorded.

Integrated Services Digital Network (ISDN): ISDN allows voice, video, and data to be transmitted over a single telephone line at speeds up to four times as fast as conventional 28.8 Kbps modems (up to 128 kilobits per second or Kbps), which many personal computer users owned in 1997 (see also ADSL). Since Spring 1998, however, 56.6 Kbps modems are becoming standard for most personal computers and can be used without requiring ISDN, thereby saving the customer telecommunication charges and reducing the ISDN advantage further (see also Table 3.1). Ultimately, ISDN is of little use if the Internet service provider's system is overloaded and/or connections to the Internet are slow (e.g., AOL in 1996/1997). The speed offered by ISDN cannot necessarily be taken advantage of in such a case.

Internet: This is a network of computer networks. The Internet makes it possible to download World Wide Web information and to receive and send e-mail. It is open to the public, thereby permitting all users to access the information on a server (e.g., a firm's Web site). In turn, this also increases the security challenges to prevent hackers and cyberpunks from unauthorized access (cf. Tables 4.1 and 4.9; see also Extranet and Intranet).

Internet Communities: (see Virtual Communities).

Internet Literacy: This could be defined as (1) an information technology user who has knowledge about computers (declarative knowledge about how the technology works), and has knowledge about how to do various tasks using a computer (procedural knowledge, i.e., how to design a database or use a communication package; Gattiker, 1990b, p. 231); (2) the user is able to use internal and external means for communicating electronically (e.g., electronic mail), to obtain information through list servers/electronic newsletters, and to use the World Wide Web (WWW) to cruise on the net; (3) the user has developed some automatic mechanisms (i.e., has limited cognitive resources that are required to do certain tasks) when using information technology (e.g., working with a local area network or a database) and the Information Highway (e.g., using file-transfer protocol to download/upload files remotely). There may be various degrees of literacy.

Internet Phone: Works similar to the telephone, but is slower (delay one second). It is substantially cheaper than an actual phone call, especially from countries with regulated markets (e.g., Switzerland and South Africa).

Intranet: An Intranet is a private computer network that uses the technology of the Internet (e.g., some browser or Web software/technology) to disseminate information within an organization. The key concept here is privacy and security: Intranets are off-limits unless one has the proper authorization. For instance, a department's server may be accessible to department employees only, without permitting them to receive and send mail from and to the Internet. Nevertheless, the employee may have access to another server and software to take advantage of the Internet (see earlier) or the Extranet (see Extranet).

IRC: Internet Relay Chat, a CB simulator that lets you have live keyboard chats with people around the world.

IS Gainsharing Plan: This is defined as an organizationwide pay and reward program, designed to reward employees for improvements in safety, security, privacy, access, and workers' skills against threats and vulnerabilities of IS, information, data, and people, using a cost-benefit approach.

Key Recovery (sometimes called *Key Escrow*): This provides some form of access to plain text outside the normal channel of encryption or decryption for a third party such as a law enforcement agency.

> *Trusted Third Party Encryption* (TTPE): Private keys are either stored with a public or private agency acting in a trust capacity. The existence of a highly sensitive secret key or collection of many keys must be secured for an extended period of time (see Key Recovery).

Latent Stakeholders: They have usually one attribute, for example, power, legitimacy, or urgency. They receive little attention by management because this group is likely to take a passive stance toward the provider of medical information (PMI).

> *Dormant Stakeholders:* They have power to impose will on the firm but may lack a legitimate relationship or urgency to do anything, for instance, media writing a story about how medical records are being used (see Latent Stakeholders).

> *Demanding Stakeholders:* The sole relevant attribute is urgency, such as a sole demonstrator calling for a particular action, for instance, a sole citizen who demands the Privacy Commissioner or ombudsperson do something (see Latent Stakeholders).

> *Discretionary Stakeholders:* They have the legitimacy attribute but no power or urgent claims on the organization, such as the cancer foundation obtaining funds and voluntary labor from the PMI or a medical insurer. They may receive help with having their database about donors and/or members put together by an employee volunteering time and know-how (see Latent Stakeholders).

Local-Area Network (LAN): This is a network of computers usually within the same room or building (see also Wide-Area Network).

Netiquette: This refers to a loose and idiosyncratic collection of rules of conduct and behavior. Life is full of etiquette and norms for various situations, such as how to behave when eating at a fancy restaurant, or when conducting a business meeting. Similarly, netiquette tries to institutionalize and make users aware of the ropes and what type of behavior may be acceptable and what may not. Consequences are informal and one might receive a nasty response if one violates netiquette.

Outsourcing: Offers companies opportunities to outsource certain production and service functions in order to reduce costs. A possible disadvantage is that scheduling and quality control become more difficult. Trust between supplier and client is a must. Recently, some firms have begun to insource some tasks again to avoid the above difficulties while, most importantly, giving their workers the chance to compete for such contracts (e.g., outsourcing vs. insourcing of software, hardware, and web page maintenance).

People versus End-Users: The term people in this context encompasses legitimate (authorized employees) and nonlegitimate users (hackers). End-users are made up of legitimate system users only, either on the Intranet, the Extranets (e.g., dealers), and the Internet (e.g., firm's web page).

Person–Computer Interaction: The more traditional term used is "man"–computer interface, which has been replaced here with a gender-neutral term. Person–computer interface could be defined as encompassing the critical factors for success to be considered before and during the acquisition of new technology and making the necessary adaptations leading to organizational change (e.g., work flow, job-related tasks, and organizational structure). Thus, person–computer interface looks at the physical context of work (e.g., work station design, illumination, ambient temperature, privacy and social interaction, and also visual and acoustical privacy) as well as ergonomics [e.g., ergonomics of hardware such as the visual display terminal (VDT) and keyboard design, software ergonomics, and user-friendliness of technology applications]. The difference between ergonomics and person–computer interface is that whereas the former looks at the physical and sensorimotor aspects, person–computer interface goes much further by considering ergonomics issues in the larger context of the work flow, work design, and the work environment in general as well as the individual's attentional and cognitive resources (e.g., knowledge, reasoning, and information processing). Hence, successful person–computer interaction leads to a healthy work environment and supports the employee's and the firm's efforts toward a high quality of work life for technology users.

"Phone Phreak": This is a person who uses technology's weakness and the phone switching equipment's vulnerability to chat with others (preferably long distance) at somebody else's expense by circumventing security measures with rudimentary technology.

PMI (Provider of Medical Information)*:* This could be a firm doing this job on behalf of the government or a government agency/department providing the service.

Pretty Good Privacy (PGP)*:* A widely available software package permitting users to use encryption when exchanging messages. Export versions of the PGP software are different than versions used in the United States and Canada.

Privacy: This may be defined as the individual's right to determine his or her own communication contacts and the right to control the use of personal information by others (Gattiker, Kelb, et al., 1997, p. 606). Additionally, it should be made technically and economically feasible for the individual and for commercial organizations to control and protect their own private data to an extent that they determine themselves and, as importantly, with measures selected at their own discretion.

Risk: In general terms, we assume that risk entails a probability that something positive or negative might occur. First we need to distinguish between two types of risks:

> *Systematic Risk:* This type of risk can usually be controlled for and is reduced by, for instance, building investment portfolios holding stocks, bonds, treasury bills, and other investment instruments—to reduce the potential for losses due to economic cycles. The risk for power outages may be managed by having a backup generator, and by limiting access to confidential data to a group of people and/or machines that absolutely need that access to perform their tasks and duties (see Risk).

> *Unsystematic Risk:* This is risk from which an investor (or information system specialist) cannot usually be protected (e.g., natural disasters looming in the future; see Risk).

Systematic and unsystematic risk together make up 1.0, the total risk. We need to assess the probability of a risk actually materializing. For simplicity's sake we call it *Probability of Risk* (PR). Accordingly, we can assess what the PR that a particular unsystematic event will occur is and what the costs culminating out of this occurrence are.

SaveNet: This is similar to a Freenet as far as being not-for-profit and supported by volunteers' time as well as equipment donations made by individuals and organizations. To avoid the creation of an "entitlement" mentality by voters, governments are encouraged to limit dollar support per user for each year (approximately $10.00 at 1995 prices) over a 5-year period. The formation of SaveNets should be encouraged primarily in communities with less than 40,000 people in order to facilitate economic access and to obtain economies of scale through a larger number of users than would otherwise be possible. The SaveNet we envision requires users to pay based on the type of service chosen (cf. Table 3.3, SaveNet). Ultimately, the SaveNet objective is simply to jump-start the use of the Internet in smaller communities, thereby achieving the economies of scale that, in turn, encourage private suppliers to provide various services at highly competitive prices.

Skills: Using the relational phenomenon, as utilized by Weberian theory, skills can be categorized using their potential ease of transferability (e.g., to another job and/or employer). Gattiker (1990a, chap. 12; 1990c) proposed such a categorization, and suggested that transferability of skills decreases from *basic* (reading, writing, and arithmetic), to *social* (e.g., interpersonal skills and the person's ability to organize his or her own effort and task performance, and possibly that of his or her peers and subordinates), to *conceptual* (including planning, assessing, decision making about task- and people-related issues, and judging or assessing tasks done by self or others), to *technology* (encompasses appropriate use of technology, such as a computer, thereby preventing breakdowns or accidents), to *technical* (physical ability to transform an object or item of information into something different), and finally, to a person's *task skills* (usually job specific).

Sociotechnical System Approach: This approach takes into consideration social issues (e.g., interaction between users, decision making and structure of groups) as well as technical ones (e.g., type of technology, ergonomics, performance, and software) when addressing the application of technology at work and during leisure time. In conjunction with the communication process, it reveals information about who communicates with whom, closeness between members of the network, as well as communication style and semantics used, including software and hardware offered by the CIS (Gattiker, 1992b, pp. 296–298).

Technology Attitudes: An attitude is generally seen as a disposition to respond in a *favorable* or *unfavorable manner* to an object (Oskamp, 1977, pp. 2–12).

Telework: Describes individuals who may be working from a satellite office and/or from home, or from another location other than the main office for part or all of their working hours (Gattiker & Hedehus, 1999). An example might be an executive who works from home on Fridays to get the requisite peace and quiet to work on correspondence, reading important reports and being able to do some conceptual work without being constantly inter-rupted (see also Virtual Organization below).

Telework can be defined as work performed with the help of com-puter-based communication technology. Accordingly, it does not imply that the employee has a constant online connection with his or her main of-fice but, instead, it may simply mean that one uses computer technology for performing tasks, while being geographically in a different location than the head office (Gattiker & Hedehus, 1999).

Telnet: Access to databases, computerized library card catalogues, weather reports, and other information services, as well as live, online games (e.g., playing bridge) that let you compete with players from around the world (see also SaveNet).

TEMPEST: Transient Electro Magnetic Pulse Emanation Standards that, if implemented on a particular hardware, prevent a party to do data snooping without the harmed party being aware of its privacy being violated.

Threat: A threat is the potential for having accidentally or deliberately compromised the availability, confidentiality, and integrity of a system or database. For instance, data may be released to other parties who should not have access to the information. Moreover, data may be compromised by having it altered, after which time it is no longer authentic. A threat can be magnified by the vulnerabilities with which a system may be faced (e.g., us-ers with limited knowledge or being careless with passwords). Accord-ingly, a system's vulnerability may increase the likelihood of a threat materializing and may thus result in a compromised information system.

Unix Box: A type of operating system that runs on most computers, similar to DOS. It is a multiuser, multitasking operating system, in comparison to DOS that is a single user, single-task system. Most of the Internet is run un-der Unix-based systems.

Virtual Communities: These communities are based on specific instances or styles of imagining community as experienced by members of the virtual space on the Internet. However, fragmentation even within a virtual com-munity of people on house remodeling may occur because one group might be interested in renovating historical houses versus another subgroup reno-vating their respective bungalows. Virtual communities may share lan-guage and culture if their membership is limited to a geographical region but in many cases, neither language nor culture nor having played the same games as children and watched the same shows may apply. Accordingly, people's interests (e.g., profession and hobbies) may support an imagination of communities in cyberspace (cf. Kjaerulff, 1998; see also communities and institutionalization).

Virtual Organization: This defines a firm that may take on different forms. For instance, an electronic shopping mall or a bookstore may be virtual by not having a retail outlet (i.e., storefront at the main shopping plaza); nonetheless, its offices may be at somebody's home whereas servers may be located (outsourced) at the premises of an Internet service provider. This type of organization may also exist by providing a virtual link between various independent small firms that, in turn, portray a unified image to the outside world (e.g., consulting services, whereby several consultants with various specialties offer a comprehensive service via the Internet). This exemplifies a social network of organizations that may culminate in a virtual firm existing over a period of time, or simply to obtain one particular contract.

The virtual organization, therefore has a

1. *physical/spatial dimension* such as location (e.g., home, office, country, and/or outsourcing of warehousing/logistics to others), human and other assets;

2. *time dimension* as represented in the firm's form that could be semipermanent or for a limited time to achieve a specific goal such as securing and completing an awarded contract; and

3. *social/organizational network dimension* exemplified by the routinized pattern of social relations among individuals, groups or teams securing and maintaining the adequate level of economic and other resources.

Virus: A program that searches out other programs and infects them by embedding a copy of itself in them, so that they become Trojan Horses. When these programs are executed, the embedded virus is executed as well, thereby propagating the "infection." This process tends to be invisible to the user.

Encrypting Virus. These viruses disguise their intent by performing some form of encryption, such as typing mga while fka appears. The decrypting code shifts the bits to reveal the executable code. This type of virus is fairly easy to be detected by most antivirus program.

Polymorphic Virus. The simplest type inserts random lines into the decryption process to throw off the antivirus scanners. Some also execute code fragments that look normal, for example, meaningful "decoy" commands are used to conceal decryption. It may even change the appearance of the actual decoding algorithm—the basic looking device. Worst is that there are usually few constant commands in the decryptor, further confounding less sophisticated antivirus programs.

Trojan (or *Trojan Horse*). A malicious, security-breaking program that is designed as something benign, such as a directory lister, archiver, game, or program to find and destroy viruses. A Trojan is similar to a back door.

Macro Virus. Macro viruses spread by having one or more macros in a document. Opening or closing the document or any activity that invokes the viral macros, activates the virus. When the macro is activated, it copies itself and any other macros it needs, sometimes to the global macro file NORMAL.DOT. If they are stored in NORMAL.DOT, they are available in all open documents.

Virtual Server: This represents a machine hosting web pages (sometimes also called a Web hotel) running on Unix and/or Windows NT whereby clients either rent server space or else place their own server at the premises of a firm specializing in virtual servers. Accordingly, the client may update and change the web pages remotely if preferred or otherwise, the company specializing in servicing/managing virtual servers will do it on the client's behalf. The advantage for the client is that telecommunication charges, software costs can be reduced. Most importantly, having the technical personnel to run the server 24 hours a day for 365 days is not necessary with the virtual server solution because this service is offered by the hosting company and is included in the rental price for the virtual server. Some estimates suggest that running a server may cost an organization easily around $100,000 U.S. per annum (e.g., hardware, software, telecommunication charges, and system personnel). Using a virtual server solution including space on a virtual server, technical support, software, and telecommunication charges may cost little more than about $1,000 per year. Naturally, if the client chooses to expand Web activities sometime in the future, these costs may be of little consequence and owning a server that is located in-house may be preferable. Nevertheless, for small and medium-sized firms, a virtual server is an economic and viable option to secure a Web presence with limited resources.

Virus and Damages: see Damages.

Vulnerabilities: According to Fahs (1997, p. 9) vulnerability can be defined as "a weakness or lack of controls that would allow or facilitate a threat actuation against a specific asset ..." or property. Risk management sometimes assumes that vulnerabilities (attitudinal and environmental attributes) could be described as an inadequacy, related to security that could permit a threat to cause harm.

WAIS: Wide-Area Information Server is a program that can search dozens of databases in one search.

Web: This term implies either using a graphical software browser such as Microsoft Explorer or Netscape to answer e-mail or using such free e-mail services as Hotmail. With such software, the user is able to visit a firm's Web site to obtain information, play games if any are offered, and place an order if so desired. The graphical interface enables the firm to take advantage of better graphics and design features for its virtual shopping window. Moreover, based on information entered by the customer about such a product as a chest of drawers, a program can simulate the looks of the product giving the customer a visual image. Subsequently, the customer may then place the order based on the specifications previously entered into the system (e.g., measurements, type of material used, and desired delivery date). The use of this technology requires, however, that the user is online while surfing the Web and has the adequate band width needed to download the information quickly (e.g., virtual simulation). However, fast downloading might require a cable or an ADSL modem.

Web Portal: This represents a heavy traffic site on the Internet with extensive content and hyperlinks. With the latter, users can connect to other sites that might have content of interest to them. The idea of a web portal is to become the starting point for a user when logging onto the Web with a graphical browser such as Netscape or Microsoft Explorer. The user begins the journey on the Web by connecting to his or her favorite Web portal. In turn, the latter usually attracts advertisers that want to target the consumers or clients visiting this site, especially if the site has areas that are of interest to particular groups of users such as a site on growing flowers in one's garden including a chat and advice forum dealing with fertilizer, seed, and pesticide issues. Accordingly, such megasites may be in direct competition for content providers such as AOL. But their services may be for free and paid by advertisers and purchases made by visitors, whereas services such as AOL also ask for a monthly fee.

Wide Area Network: Connects several LANs (see Local Area Network) or other wide area networks to each other; the Internet is a huge wide area network.

WWW: World Wide Web was developed by researchers at the European Particle Physics Laboratory in Geneva; the WWW is somewhat similar to the WAIS. However, it is designed on a system known as hypertext; that is, words in one document are "linked" to other documents. It is sort of like sitting with an encyclopedia—you read an article, see a reference that intrigues you, and so flip the pages to look up that reference.

Appendix B

Technology and Innovation Management (TIM)-Research

The Center for Technology Studies at The University of Lethbridge (Canada) and Aalborg University (Denmark) as well as the Corporate Technology and Environmental Management Group at the Aarhus School of Business (Denmark) have been collaborating on various Internet projects since 1995.

Current projects include:

- Assessing end-user behaviors on the Internet
- Privacy and security management
- World Wide Web and customer relations
- Institutional environmental reporting and the World Wide Web
- "Malware" and communication/data exchange
- Reengineering and benchmarking

A survey is currently being prepared to gauge the needs of business and private users with respect to safety/security, privacy, and marketing issues. If you are interested in participating in this survey or wish to receive more information about past project findings or references for this article, please contact:

E-Mail: Internet_Research_Program@bigfoot.com
Center for Technology Studies
Department of Production

Aalborg University
Fibigerstraede 16
9220 Aalborg O
DENMARK
Web: http://research.WebUrb.dk

Material in this Appendix is further addressed in chapters 6 and 7.

ETHICAL AND MORAL CONCERNS ON THE INTERNET

Following are the vignettes used in Phase 1 of the research program described to some extent in chapters 2 and 5 of this book. Vignette 2 lists the questions (probes) that were asked of each participant. Vignettes 1 and 3 use the same probes except for slight modifications for fitting questions to each scenario.

Situation 1: Data Encryption

One of your friends is a technical whiz and has just developed a new data encryption device (i.e., similar to a phone scrambler); the device helps to protect conversations against wiretapping and software. Your friend quickly demonstrates how the device works by sending an encrypted message to you. Your subsequent decoding efforts fail, illustrating that the encryption device does its job very well. You and your friend then proceed to install this device and software on both of your machines for utilization when communicating with each other.

Situation 2: Virus

One of your friends is a real computer nut and has just written a new computer virus program. Your friend then proceeds to load the virus program onto a BB or an electronic newsletter/listserver (EDL).

What do you think about this situation (posting a computer virus one wrote onto a BB or an EDL)?

Very wrong

A little wrong

Perfectly okay

Is anyone hurt by what your friend did?

No   Yes

If Yes, Who/Person?

How?

Imagine that you actually saw someone posting a virus program he or she wrote onto a BB or an EDL. Would you:

feel bothered

not care

think this is good

Should the person be stopped?   No   Yes

Should the person be punished?   No   Yes

Suppose you learn about two different countries. In country A, people posting design or programming characteristics of a computer virus they wrote on a BBS or an EDL are quite common, and in country B, one never shares such programming information about a virus one has created with members of a BBS or an EDL.

Which one of these customs (if either) is bad or wrong?

Both customs are wrong

Country A's custom is wrong

Country B's custom is wrong

Neither one, both customs are okay

Situation 3: Illegal Computer Game

Your friend has just received a new computer game through an EDL located abroad. The game is banned in this country because of its violence, and its sexual and racist content. Your friend tests the game. Although he or she finds it somewhat disgusting, your friend sends a copy to another friend abroad, where no regulation exists possibly banning the game. Your friend does not keep a copy of the game.

COMMERCIAL USE OF THE INTERNET: PRIVACY AND SECURITY—PHASE 2

Following are the vignettes also used in chapter 6 (Phase 2 of the project) of this book. Vignette 2 lists the questions (probes) that were asked of each participant. Vignettes 1 and 3 use the same probe except for the two questions dealing with the 50¢ payment, and slight modifications to fit the questions to each scenario (e.g., instead of advertising, using the word monitoring in Vignette 3).

Situation 1: Caller Number Identification (CNID)

A month ago your friend placed a fast food order. Thanks to Caller ID, Mamamia Pizza was able to call back to confirm the order and the street address just as a precaution against possible pranksters. This week your friend received a phone call from Carleone's Delight (another pizza chain) informing him or her about their special for new customers. After some prodding, your friend was told that Carleone's had purchased the number from a telemarketing company (which in turn purchased the information courtesy of Mamamia's database), including information about his or her habit of ordering in pizza about 3–4 times a month. Your friend always uses per-call blocking (i.e., his or her number is blocked by dialing *67 before calling any retailer) but he or she must have forgotten when calling Mamamia last time.... Today, "Uncle Charlie" called with their special ..., again thanks to Mamamia's database....

Situation 2: Electronic Advertising

Your friend pays a fixed U.S. $20.00 fee each month for Internet access regardless of use. As an avid traveler, he or she is also on the AdventureTravel list. Your friend is very concerned about his or her privacy but as a newcomer to the Internet is not aware that email addresses are not concealed on the Listserver managing the electronic mailing list unless the user specifically requests this. Logging on to the system today reveals that two commercial messages are in the "inbox"; every time your friend reads such a message, a 50¢ CREDIT is given to his or her account. Your friend can choose to delete unread advertising messages when logging out. The first advertising message is from the CEO of Kelley's Cruises addressed to "Dear Fellow AdventureTravel netter ..." informing him or her about the special for AdventureTravel netters. Your friend then inquires to the sender about how Kelley's Cruises got his or her email address. Answer: Kelley's CEO is a fellow AdventureTravel netter. Because your friend's email address was not concealed on the list, it and other email addresses on the list were obtained using a simple command.

1a. How do you feel about using "notconcealed" email addresses of electronic mailing list subscribers for an advertising campaign?

> Very wrong
>
> A little wrong
>
> Perfectly okay

How do you feel if users are paid 50¢ for reading each message?

> Very wrong
>
> A little wrong
>
> Perfectly okay

1b. Is anyone harmed by what Kelley's Cruises did?

> No   Yes
>
> If yes, how?

1c. Imagine that a retailer suddenly sends you electronic advertising based on your "notconcealed" email address obtained from an electronic mailing list. Would you:

> feel bothered
>
> not care
>
> think this is good

How would you feel if you were paid 50¢ for reading the message?

> feel bothered
>
> not care
>
> think this is good

1d. Imagine that you are a Kelley's Cruises employee sending an individual an advertisement on the Internet/Information Highway, based on the person's "notconcealed" email address obtained through an electronic mailing list. Would you:

feel bothered

not care

think this is good

How would you feel if the recipient was paid 50¢ for reading the message?

feel bothered

not care

think this is good

1e. Should Kelley's Cruises be stopped?

No   Yes

If yes, how?

1f. Should Kelley's Cruises be penalized?

No   Yes

If yes, how?

1g. Should the government intervene?

No   Yes

If yes, how?

Suppose you learn about two different countries. In country A, using membership lists from discussion lists and electronic mailing lists for commercial purposes, such as advertising, is a common practice (one can, however, choose not to read the message sent or else get a credit to one's account). In country B, legislation exists such that consent to the use of personal information (including one's not concealed email address on a discussion list or electronic mailing list) must be manifest, free, enlightened, and given for specific purposes by the individual to the firm or other user; however, this consent is valid only for the length of time necessary to achieve the purposes for which it was requested (i.e., Kelley's CEO would have had to ask you if he or she could use your email address BEFORE doing so).

1h. Which one of these customary practices (if either) is bad or wrong?

Both customary practices are wrong

Country A's customary practice is wrong

Country B's customary practice is wrong

Neither one is wrong, both customary practices are acceptable

1i. Do you personally know of anyone who has ever received advertising information through email?

No   Yes

Situation 3: Privacy and Security Issues

Your friend lives in San Francisco, California, and is the administrator/moderator of an international mailing list called "Future Tech." Your friend runs this list from the PC at his residence over a designated telephone line. This list discusses hightech issues, such as new product developments, marketing on the Internet/Information Highway, strategic and precision weapons, systems integration, defense electronics, and new developments in commercial and military aerostructures. Your friend uses this network to communicate regularly with associates in locations around the globe. Your friend finds out that:

1. Stating national security reasons (e.g., possible product espionage), U.S. security agencies have obtained a court order to "tap" your friend's email account, permitting the agencies to monitor both incoming and outgoing electronic messages between the system administrator (your friend) and mailing list members.

2. Having received court authorization, the U.S. security agencies have requested the telephone company to give it realtime access to your friend's transaction information (i.e., obtaining "call setup information" for "live" monitoring of incoming and outgoing electronic messages on the mailing list).

Your friend knows from officials of the agencies that the information obtained from the wiretaps is being used to build databases on foreign members of the "Future Tech" network and on hightech developments that have implications for "national security." Your friend's computer buddies have complained to their countries' respective representatives from Information/Privacy Protection agencies, claiming that the U.S. security agencies are collecting information originating outside U.S. borders where the U.S. court order may not apply. Hence, the privacy of foreign nationals is violated. Moreover, because your San Francisco friend has not been charged with any illegal activity by the California State court or the Federal court, his or her foreign buddies are complaining that the wiretapping may violate privacy under U.S. laws.

E-COMMERCE, MARKETING, AND PROFITABILITY—PHASE 3A

This Phase of the project and the issues to be addressed are discussed quite extensively in chapter 7.

NETWORK MANAGEMENT, MARKETING, AND PRODUCTIVITY—PHASE 3B

Following are the vignettes used in Phase 3b of the project which is currently under progress. Vignette 1 lists the questions (probes) that were asked of each participant. Vignette 2 uses the same probe except for the last two questions as well as slight modifications for fitting questions to each scenario.

Situation 1: Monitoring Internet Use at Work

Your friend has been using his or her employer's Internet facilities for several months. The company has a policy that use of the Internet should be limited to jobrelated purposes only. The company also reserves the right to monitor, at random, traffic on the company's network. Last month your friend decided to do some personal banking from the workstation at the office, after work hours, to pay some bills via his bank's Internet/Web facilities. This month, for the first time, the employer sent him or her a bill in the amount of $1.50 for the 30-minute personal use of Internet access. The company provided your friend with the time, duration, and web sites (URLs) he or she accessed during the previous month, including the time during which the bank transfer occurred.

1a. What do you think about your friend's Internet activities being monitored by his employer?

Very wrong

A little wrong

Perfectly okay

1b. Is anyone harmed by what the employer did?

No   Yes   Maybe

Why not?

If yes, how?

Comments?

1c. What do you think about the employer charging your friend for his personal Internet use?

Very wrong

A little wrong

Perfectly okay

1d. Imagine that you were in a situation where your employer monitored Internet access of all its employees. Would you feel:

bothered

not care

think this is good

1e. Imagine that your employer asked you to monitor Internet access of all employees. Would you feel:

bothered

not care

think this is good

1f. Should the employer be stopped from monitoring Internet access?

No   Yes   Maybe

Why not?

If yes, how?

Comments?

1g. Should the employer be stopped from charging for personal Internet activities?

No   Yes   Maybe

Why not?

If yes, how?

Comments?

1h. Should the employer be penalized for monitoring?

No   Yes   Maybe

Why not?

If yes, how?

Comments?

1i. Should the government intervene?

No   Yes   Maybe

Why not?

If yes, how?

Comments?

Suppose you learn about two different countries. In country A, firms that monitor their employees' activities are quite common. In country B, legislation exists such that an employer is not permitted to monitor their employees' computer usage due to privacy laws that apply at the workplace.

1j. Which one of these practices (if either) is bad or wrong?

Both customary practices are wrong

Country A's customary practice is wrong

Country B's customary practice is wrong

Neither one is wrong, both customary practices are acceptable

1k. Do you personally know of anyone whose Internet access has been monitored by their employer?

No   Yes

11. Do you personally know of anyone who has been charged for nonwork, that is, private Internet use by the employer?

                                                          No   Yes

Situation 2: Privacy and Security in Commerce

Your friend is "visiting" a large insurance firm's Internet/Web site and browses through the firm's information about its insurance services. Accident insurance covering the risk incurred of such sports as bungee jumping, mountain climbing, sky diving, and so forth, is described with very attractive rates, available only to clients on the Internet. Because your friend regularly goes sky diving, he or she decides to sign up for this insurance. Some basic information has to be provided (e.g., age, gender, marital status, address, profession, email address) as well as basic data about one's health and any previous accidents one may have had. Your friend is approved and receives the insurance policy by mail within 5 days.

A month later your friend tries to get a mortgage to finance the purchase of his or her dream house. But at the bank, he or she is informed that due to participating in "high risk" sports, the bank requires your friend to take out an additional insurance policy that will be used to pay off the mortgage in case of serious accident or death. Your friend is surprised that the bank knows about his or her sport habits because he or she was not asked to provide such information anywhere on the loan application.

After inquiring with the loan officer, your friend is told that the bank owns the insurance firm that offered your friend insurance covering the sky diving risk and that the group of companies uses one customer database (insurance policy and loan data).

REFERENCES

For additional information on the research program, please see the list of publications below.

Gattiker, U. E. (December, 1994). Im Cyberspace-Café. Bräuche und Missbräuche auf dem Information-Highway (The Cyberspace coffee-shop. Use and misuse on the information highway). *Tüte - Tübinger Termine (Special Edition on Daniel Foucault)*, 46–50.

Gattiker, U. E., & Kelley, H. (1994). Techno-crime and terror against tomorrow's organisation: What about cyberpunks. E. Raubold and K. Brunnstein (Eds),

Proceedings of the 13 World Computer Congress - IFIP Congress '94, Hamburg (pp. 233–240). Amsterdam: Elsevier Science Publishers.

Gattiker, U. E., & Kelley, H. (1995). Morality and technology, or is it wrong to create and let loose a computer virus. In J. F. Nunamaker, Jr. & R. H. Sprague (Eds.), *Proceedings of the 28th Annual Hawaii International Conference on System Sciences 1995* (pp. 563–572). Las Alamitos, CA: IEEE Computer Society Press.

Gattiker, U. E., & Kelley, L. (1999). Morality and computers: Attitudes and differences in moral judgments across populations. *Information Systems Research*, 10, 450–475.

Gattiker, U. E., Holsten, H., & Miller, J. (2000). Attitudes of users toward surveillance on the internet: A South African field study. In U. E. Gattiker, P. Pedersen & K. Petersen (Eds.), *Conference Proceedings EICAR International Conference*, ISBN: 87–997271–0–9.

Gattiker, U. E., Janz, L., & Schollmeyer, M. (1996). The Internet: What are some of the benefits and costs for your organization. *Business Quarterly*, *61*(1).

Gattiker, U. E., Kelb, J., & Miller, J. (1999). Direct marketing and privacy for telephone and internet users. Working Paper, submitted for publication.

Gattiker, U. E., Kelley, H & Janz, L. (1995). Morality, computers and the Internet: Managing a LAN. *Proceedings of the 95 EICAR Conference - Improving Security of PCs and PC-Networks* (pp. 120–151).

Gattiker, U. E., Kelley, H., & Janz, L. (1996). The Information Highway: Opportunities and Challenges for Organizations. In R. Berndt (Ed.), *Global management* (pp. 417–453). Berlin & New York: Springer-Verlag.

Gattiker, U. E., Janz, L., Kelley, H. & Schollmeyer, M. (1996). The Internet and Privacy: Do you know who's watching? *Business Quarterly*, *60*(4), 1–6.

Gattiker, U. E., Perlusz, S., Bohmann, K., Seiferheld, I., & Ulhøi, J. P. (1999). How corporate clients and consumers surf the Internet: A review and future directions for research. *Proceedings of European Computer Information Systems Conference 99*, Copenhagen (pp. 212–234).

Gattiker, U. E., Janz, L., Greshake, J, Kelb, J., Schwenteck, O., & Holsten, H. (1996). Internet and Organisations: Social Aspects of Information Security. In C. Schmid (Ed.), *Proceedings of the 96 European Institute for Computer Anti-Virus Research (EICAR) Annual Conference, Linz, Austria — Malicious software and the Internet* (pp. 185–203).

Gattiker, U. E., Schwenteck, O, Greshake, J., Janz, L., Kelb, J., & Holsten, H. (1997). The Internet community and ethics: A cross-national field study. *Proceedings of the Administrative Science Association of Canada - Technology and Innovation Management Division*, *12*/Part 7, 77–86.

Gattiker, U. E., Kelb, J., Janz, L., Holsten, H., Greshake, J., Schwenteck, O., & Miller, J. (1997). Direct marketing and privacy for telephone and Internet users: A South African field study. *Global business in practice. Proceed-*

ings of the Tenth International Bled Electronic Commerce Conference, Bled, Slovenia, 604–639.

Gattiker, U. E., Fahs, R., Blaha, J., & members of EICAR Working Group 1. (2000). Managing medical information systems: Can patients' privacy be protected or should we simply give up? *International Journal of Healthcare Technology and Management.*

Appendix C

Telephone Marketing

Computer Privacy Digest Sat, 11 Feb 95
Volume 6 : Issue 017
From: "Mich Kabay [NCSA Sys_Op]" <75300.3232@compuserve.com>
Date: 10 Feb 95 07:42:31 EST
Subject: Autodialing Ban

From the Associated Press news wire via CompuServe's Executive News Service:
Apn 02/06 2020 Autodialing Ban
By BOB EGELKO
Associated Press Writer

SAN FRANCISCO (AP)—A federal ban on automated, tape-recorded telephone sales pitches was upheld Monday by a federal appeals court, which said Congress accurately identified commercial autodialing as a threat to privacy.

Key points made by the author:

- The original ban was suspended in December 1992.
- The new ruling validates the law "prohibiting use of automatic dialing machines to reach homes unless the consumer has consented to get such calls or the message is introduced by a live operator."

- The National Association of Telecomputer Operators challenged the original law, claiming it discriminated against small commercial organizations who cannot afford live telemarketing operators.
- The law in question also bans automatic junk-faxes.
- Congressional testimony at the time of passage of the legislation described automated calls as

... more invasive than live calls because the taped messages "cannot interact with the customer except in preprogrammed ways" and [they] "do not allow the caller to feel the frustration of the called party."

- Such calls, according to the judge, "also cluttered answering machines and made it difficult for consumers to remove their names from calling lists."

M. E. Kabay, PhD., Director of Education, National Computer Security Association (Carlisle, PA); Mgmt Consultant, LGS Group Inc. (Montreal, QC)

End of Computer Privacy Digest V6 #017

Appendix D

How Does Caller ID Work: Privacy at Large

Caller ID (or also called CNID) in the United States, or Caller Display as it is known in England, costs about U.S. $2.00 or more per month in North America or less in other countries such as England. In Canada, the subscriber pays about U.S. $5.00 per month for the service (e.g., Province of Alberta) and receives the phone number of the caller as well as the name of the person calling (i.e., name of the subscriber or individual who is listed on the monthly phone bill sent to the address from where the call is originating).

In Canada, some provinces offer the individual telephone subscriber the option to block out the entire line of caller data, that is, the number is not revealed to any user, not even to the police when calling in an emergency. In contrast, British Telecom (BT) will block entire lines but only after their approval (e.g., for women shelters). Customers can block Caller Display/ID by dialing 141 before each call. However, BT claims that over 70% of customers "see no occasion where they might need" to use the 141 feature.

Alternate Number Display (AND) allows a number unique to the customer and different from one's phone number to show up on a called party's Caller ID box. The number cannot be called back and anyone who tries will get a message to the effect of, "The party you are trying to reach does not accept calls at this number."

In 1994, British Telecom introduced Call Return; customers dial 1471 and (unlike in the United States) will hear the phone number of the person who called them last. The service is free.

In 1995, NYNEX (local phone company for New York and areas of New Jersey) had difficulty with this technology, because Caller ID permitted individuals to receive phone numbers of callers who had their phone numbers unlisted (i.e., not listed in phone directory or available through directory assistance). Privacy advocates were outraged and NYNEX was forced to address this issue promptly to the satisfaction of phone subscribers with unlisted phone numbers.

Appendix E

No Privacy for Corporate E-Mail?

From epicnews@epic.org Thu Feb 29 20:53:03 1996, Volume 3.05, February 29, 1996
Published by the Electronic Privacy Information Center (EPIC) Washington, D.C. E-Mail: info@epic.org; http://www.epic.org/

[4] Court Rules Against E-mail Privacy
A U.S. District Court in Pennsylvania ruled on January 18 that even if an employer promises not to intercept e-mail on a company system, there is not an expectation of privacy in the e-mail. The case underscores the question of whether the court system has an adequate understanding of the underlying technology.

According to the court, the employer repeatedly told its employees that "all e-mail communications would remain confidential and privileged" and that "email communications could not be used by [the company] against its employees as grounds for termination." Even with that promise, the employer intercepted the communications of Michael Smyth and fired him. The court held that:

> We do not find a reasonable expectation of privacy in email communications voluntarily made by an employee to his supervisor over the company email system notwithstanding any assurances by management. Once plaintiff communicated the alleged unprofessional comments to a second person over an email system which was apparently utilized by the entire company, any

reasonable expectation of privacy was lost ... we find no privacy interests in such communications.

We do not find that a reasonable person would consider the defendant's interception of these communications to be a substantial and highly offensive invasion of his privacy.

A copy of the decision is available at:
   http://www.epic.org/privacy/internet/smyth_v_pillsbury.html

Appendix F

Practical Advice for Managing Privacy and Safety of Personal Information of Consumers, Employees, Students, or Other Types of Data Generated or Administered by an Organization

When an employee joins your firm, the Privacy Code of Conduct sheet below should be signed. In addition, the Code of Conduct should be administered electronically or on paper at regular intervals (cf. Screen 1–3). These procedures apply to any authorized users of your information system.

SCREEN 1

PRIVACY CODE OF CONDUCT

FOR USE OF INFORMATION SYSTEMS
AND PERSONAL INFORMATION AT

(Name of Organization/Agency)

(Rev. 2, 1997)

PREAMBLE

Dear Fellow Information System User:
The purpose of this Code is to reaffirm the organization's policy of conducting its business in full compliance with the letter and spirit of the law and the highest level of ethical standards. The Code summarizes the principles that should guide the conduct of all employees as they use the information system in the performance of their duties for this organization.

Users of this information system are subject to having their activities on this system monitored and recorded by system personnel. Unauthorized users or those persons using this system without or in excess of their authorization expressly consent to such monitoring and are advised that if such monitoring reveals possible evidence of criminal activity, system personnel may provide such evidence to law enforcement officials.

To protect individual rights, users are herewith advised that random checks of system contents by system personnel occur, thereby ensuring that all parties have accepted and are adhering to and abiding by the "Golden Rules" outlined below.

The following definitions apply in this code:

Consent—voluntary agreement with what is being done or proposed and where information of such action or proposal has been provided. Consent can be either express or implied. Express consent is given explicitly, either orally or in writing. Express consent is unequivocal and does not require any inference on the part of the organization seeking consent. Implied consent arises where consent may reasonably be inferred from the action or inaction of the individual.

Data-subject—an identifiable entity about whom or which information is processed, including but not limited to collection, recording, and storing.

Information system—the framework of standards, procedures, and databases for processing data into usable information. It is computer based and in this organization includes _____.

Personal information—information about an identifiable individual that is recorded in any form and that cannot be associated with any other particular individual.

For a client, such information includes, but is not limited to, name, address, telephone number, and information relating to that client's account with this organization.

For an employee, such information includes, but is not limited to, that employee's personal employment files, performance appraisals, career development records, and also medical and benefits information.

Process—the act of collecting, recording, retrieving, transferring, storing, or otherwise using personal information for normal business activities or as required by law.

User—an individual or individuals that obtain access to this information system to engage in collecting, recording, manipulating, retrieving, transferring, storing, or otherwise using data.

SCREEN 2

The Privacy Code of Conduct (the "Code") is intended to be followed by any users of this information system. The Code provides a comprehensive statement of principles and guidelines concerning the processing of personal information, including collection, storage, protection, use, disclosure, individual verification, and correction.

The principles[1] included in this Privacy Code of Conduct constitute a collection of acceptable modes of behavior when using the information system at or on behalf of this organization. This organization is committed to maintaining the accuracy, confidentiality, security, and privacy of data-subject personal information. In order to remain at the forefront of privacy, system safety and security, and while strengthening ethical conduct by all parties using and being affected by the use of our information system, this organization abides by the following Golden Rules:

1. Accountability

This organization (_____) is responsible for personal information under our control; _____ is designated as the person who is accountable for this organization's compliance with the following principles.

   1.1 Every action or procedure taken on any privacy matter related to processing of personal information shall be in compliance with Bill 68 (Quebec, August 1993) and the Council of European Community's Directive on Protection of Privacy (July 25, 1995).

---

[1] The principles in this Code have been adapted from the 1995 draft codes for use of personal information as proposed by the Canadian Standards Association (see also footnote 2).

2. Identifying Purposes

The purposes for which personal information is collected shall be identified by this organization before or at the time the information is collected.

2.1 Personal information on data-subjects contained in this database is used solely for the purposes of normal business activities of this organization.

2.2 Aggregate data may be used thereby making identification of individuals impossible.

3. Consent

The knowledge and consent of the individual are required for processing or disclosure of personal information except where inappropriate.

3.1 The databases of this organization are constructed, maintained, or used under the "opting in" principle. Therefore, data-subject must give informed, written consent before any other type of information can be processed. Lack of such consent implies that the data-subject has "opted out" and hence has not provided this organization with consent to process such data.

3.2 Data-subjects are not subjected to unsolicited information with regard to products or services from this organization or a related organization, except where data-subjects have provided written consent for receipt of such information.

3.3 Consent forms are retained by this organization for a minimum of 10 years.

4. Limiting Collection

The collection of personal information by verbal, electronic, or paper-based means shall be limited to that which is necessary for the purposes identified by this organization. Information shall be collected by fair and lawful means.

5. Limiting Use, Disclosure, and Retention

Personal information shall not be used or disclosed for purposes other than those for which it was collected except with the consent of the individual or as required by law. Personal information shall be retained only as long as necessary for the fulfillment of those purposes.

5.1 Personal information is not sold or transferred to any other individual or organization, unless the data-subject has provided this organization with prior written consent (cf. point 3 above).

5.2 Personal information, including information that may be on the Local-Area Network, is not transferred by tape, diskette, telecommunication lines, or any other means to any other data storage medium except for normal back-up or safety purposes for this organization, or as required by law.

6. Accuracy

Personal information shall be as accurate, complete, and up-to-date as is necessary for the purposes for which it is to be used.

7. Safeguards

Personal information shall be protected by security safeguards appropriate to the sensitivity of the information.

7.1 This organization endeavors to protect personal information against such risks as loss or unauthorized access, destruction, use, modification, or disclosure, by using appropriate security measures.

7.2 Personal information disclosed to third parties shall be protected by contractual agreements stipulating the confidentiality of the information and the purpose for which it is to be used.

8. Openness

This organization shall make readily available to individuals specific information about its policies and practices relating to its handling of personal information.

9. Individual Access

Upon request, a person shall be informed of the existence, use, and disclosure of personal information about that individual and shall be given access to that information. An individual shall be able to challenge the accuracy and completeness of the information and have it amended as appropriate.

10. Challenging Compliance

An individual shall be able to challenge compliance with the above principles with the person who is accountable with the organization.

SCREEN 3

Our firm/network abides by the above set of golden rules on data privacy. Any violation thereof results in the immediate loss of user privileges and possible legal charges for damages and will affect a worker's employment status (i.e., after first violation, individual is given written warning; second violation results in immediate termination of employment).

Employees, clients, and other users are solely responsible for their opinions, behaviors, and actions when using this database/LAN/information system and the Internet in general.

Our organization tries to exceed generally accepted and adhered to privacy standards (e.g., EU Directive on the Protection of Personal Data, 1995; Quebec Bill 68, 1993) regardless of where geographically and under which legal jurisdiction it or any of its subsidiaries and affiliates conduct business.

The system/LAN/database provider does not verify, endorse, or otherwise vouch for the contents of any note, information, or behavior on the system and cannot be held responsible in any way for the information contained or exchanged in any note or message, or for any transaction between users.[2]

I understand the 10 Golden Rules outlined above

No__ Yes__

I will abide by the 10 Golden Rules outlined above

No__ Yes__

---

[2]Of course, ultimately the firm can be held liable as far as employees are concerned when they are representing the firm in their job capacity to others, that is, job-related activities using the database/Internet/LAN).

I assume responsibility for any possible liability incurred by the database/information system/LAN/Internet access owner or provider occurring from any of my actions or behaviors that might be violating any one of the above regulations and rules

No___   Yes___

When using the database/information system/LAN/Internet, I do abide by the firm's Privacy Principles for Users of Personal Information[3]

No___   Yes___

I acknowledge that the system or network owner or provider may invade my privacy in order to safeguard the above principles for all users

No___   Yes___

I acknowledge that the system or network owner or provider may invade my privacy in order to safeguard the above principles for all data-subjects (e.g., students, employees, and customers)

No___   Yes___

I have read the above material and understand fully its implications for my use of the database/information system/LAN/Internet

No___   Yes___

I will use the database/information system/LAN/Internet strictly according to the conditions outlined above

No___   Yes___

---

[3]The firm's Privacy Principle for Users of Personal Information must be developed and must follow local laws, and data security and safety legislation must also be met (see also Appendix H, chap. 4 with Table 4.3, and also Tables 7.2 and 7.3).

If local laws are not helpful, we suggest you have a look at:

Bill 68 1993, chapter 17. (1993, August 4). An act respecting the protection of personal information in the private sector. *Gazette Officielle du Québec, 125*, 4253–4279.

Council of European Communities. (1992, November/December). Amended proposal for a council directive. *Transnational Data and Communications Report* [COM (92) 422 Final -SYN 283, October 15, 1992, pp. 32–41]. This proposal with slight modifications was approved by the Council of Ministers (July 25, 1995; European Commission Press Release: IP/95/822) and gave the 15 member states 3 years to implement the Directive.

Draft Privacy Principles. (1995, January 20). *National information infrastructure: Draft principles for providing and using personal information and commentary.* Washington, DC: Office of Management and Budget. (for information contact Jerry Gates, Privacy Working Group, Bureau of the Census, Room 2430, Bldg 3, Washington, DC 20233 or e-mail: ggates@info.census.gov; see Home Study Handout!)

As a user, I not only abide by this Code of Conduct but, most importantly, comply with it by:

a) Upholding and promoting the principles of this code.

b) Treating violations of this code as inconsistent with membership in, or doing business of any kind with, this organization.

c) Accepting one's duty to work to correct rules and norms in this Code that are wrong (e.g., according to the law and/or ethics and morals).

d) Assessing issues as addressed herein beyond my own interests but also from the collective's or society's point of view.

e) Upholding implied limits to nondiscrimination due to gender, religion, age, and others.

EMPLOYEE: WITNESS:

Name_____   Name_____
(Please Print)   (Please Print)

Signature_____   Signature_____

Date_____   Date_____

Bill 68 follows the Council of European Communities' Directive on Protection of Personal Data (approved July 25, 1995, and therefore must be implemented by all member countries by 1998) and has been law since 1993, thereby affecting the province's businesses and companies across Canada on how they handle personal data. These are good documents to look at to get an idea of what is to come in North America and around the world, considering that other Canadian Provinces, as well as some U.S. states (e.g., California), are moving in the same direction.

The Canadian Standards Association has developed a working draft for the Protection of Personal Information being discussed by various interest groups before the standard is adopted (possibly by the end of 1995):

Canadian Standards Association. (1995, May 5). *Model code for the protection of personal information* (Working draft, CAN/CSA-Q830-1995).

Also worthwhile is the following privacy code implemented by a Canadian telecommunications carrier:

AGT Limited. (1995, May 11). *Code of fair information practices.* Edmonton, Canada: The author. More information or a copy can be received from Telus Privacy Commissioner, Floor 21E, 10020 - 100 Street, Edmonton, Alberta, Canada T5J 0N5, or e-mail: menglish@ccinet.ab.ca

HOW SHOULD THE CODE OF CONDUCT BE ADMINISTERED WITH EMPLOYEES, CLIENTS, AND OTHER GROUPS OF USERS?

The above three screens should be shown, read, discussed, and signed by each employee, client, or user and a witness before providing access to the database/LAN/Internet access/system (hard copy!). Moreover, the user should be compelled to read the above material and answer the questions at least TWICE EACH 12 MONTHS. The system should then keep a record of the last 24 months (= 4 confirmations) as read and completed by the individual.

Please remember, showing these screens when logging on to the LAN/database without requiring the user to answer some questions (or filling out the form) is not enough, because the user may fail to read the material or claim ignorance later on.

Courts in various countries have been known to accept these claims by users. Hence, the user must be required to answer a few questions as listed above, as records must be kept of when one did answer the questions. Moreover, an answer indicating a "no" would automatically alert systems personnel and result in a refusal of access to the system.

SOME FINAL NOTES

The authors, Urs E. Gattiker and Linda Janz, give permission to use this document as long as acknowledgment of this document and reference to it are made and the authors are informed about the use of this code or parts thereof.

The above material is upgraded continuously and more recent releases can be obtained from Urs E. Gattiker, Department of Production, Aalborg University, Denmark. E-mail: Urs_the_Bear@bigfoot.com. Upgrades are also on the Web under: http://home.uleth.ca/mancts/infohigh.htm

Appendix G

Interesting Web Sites
for Tips, Hints, and Information

INFORMATION TECHNOLOGY: SECURITY, RISK,
AND INVESTMENT DECISIONS

The following sites provide the reader with information about how to improve security and reduce risks of unauthorized access to a CIS, risks of software piracy, and risks to privacy.

<URL:http://www.eicar.com/>
European Institute for Computer Anti-Virus Research (EICAR) with information for members and others about malicious software, viruses, firewalls, network engineering, and management; annual meeting and many other interesting hotlinks to other web sites

<URL:http://www.spa.org/piracy/homepage.htm>
Tips against software piracy

<URL:http://www.autodesk.com.au/company/nopirate>
Information about software piracy and its consequences

<URL:http://www.jou.ufl.edu/siemens/articles/0695/695naue.htm>
Security guaranteed

<URL:http://www.sni.de/public/uk_sys/whatsnew/DIA-96/firewall/firewall.htm
  Firewall

<URL:http://www.baynetworks.cz/products/network/netmgmt.html>
  Take control of your network

NETWORK MANAGEMENT SOFTWARE AGAINST POSSIBLE UNNECESSARY USE OF THE INTERNET DURING WORKING HOURS

The programs listed below enable a firm to access the Internet via an IP address; that is, one of about 2–10 computers has Internet access with a modem and telephone hookup; all other computers obtain Internet access through this machine only. These programs enable the network administrator to determine which computers will get Internet access to which Web sites (e.g., access to Playboy can be blocked). Access to Newsgroups, Internet Relay Chat, and even e-mail can be restricted to certain hours (e.g., before 9:00 a.m. and after 5:00 p.m.) or even be disabled (e.g., Relay Chat).
 The programs listed below are just a selection and all of them offer trial software to potential clients.

<URL:http://www.quarterdeck.com>
  Costs at least U.S. $ 700, works with Novell NetWare software only.

<URL:http://www.vicomtech.com/>
  The private version for one's home (2 computers) costs U.S. $99, the fully-fledged commercial version costs U.S. $399 (Macintosh and OS systems).

<URL:http://www.webster.com>
  For larger and more complex networks, minimum license is for 50+users and starts by U.S. $2,195 annually.

ERGONOMICS

The Web addresses below provide the reader with some information about how to design and implement workspace using CIS meeting ergonomic requirements and, thereby reducing absenteeism due to ergonomic-related health problems (e.g., headaches through wrong seating posture) and work-related accidents and injuries. These sites provide links to others around the world.

<URL:http://www.iha.bepr.ethz.ch/pages/ergo/bild.htm>
  designing appropriate computer workplaces

<URL:http://www.iha.bepr.ethz.ch/pages/ergo/licht.htm>
  good lighting with appropriate light

<URL:http://www.iha.bepr.ethz.ch/pages/ergo/raum.htm>
  air quality and climate/room temperatures

<URL:http://www.iha.bepr.ethz.ch/pages/ergo/sitzen.htm>
  what is an ergonomic chair?

INFORMATION TECHNOLOGY, INTERNET, AND CUSTOMERS

The sites listed below enable the customer to register the Web site and find out other pertinent information for making one's Web site customer friendly. Unless one's Web site is known, it cannot be taken advantage of by customers and potential clients (see also chap. 6, Tables 6.5 and 6.7)

<URL: http://204.57.42.244/submit.htm>
  How to register one's Web site on 18 search engines all at once

<URL: http://www.cwawe.com/bbw/subsrv.htm>
  How to register one's Web site on 30 search engines all at once

USING THE INTERNET TO LOWER LONG-DISTANCE PHONE CHARGES

Most of the Internet phones listed can be downloaded and tested for a while before the client has to decide and either pay or is no longer able to use the software. All that is needed is a modem and a soundblaster, which most new machines have anyway. The savings can be substantial, and some programs even offer video capabilities as well as good sound quality.

<URL:http://eurocall.com/d/index/.htm>
  German Internet phone

<URL:http://www.intel.com/iaweb/cpc/index.htm>
   Intel's Internet phone

<URL:http://www.web.mit.edu/network/pgpfone/>
   This software enables one to encrypt communication, free (Macintosh
   and Windows 95)

<URL:http://magenta.com/cyberphone>
   Good sound quality, simple operation (Windows and Unix)

<URL: http://www.freetel.com>
   Good sound quality, simple operation, works nicely with 14.4 modem,
   simple version without the bells and whistles can be used for free (Windows).

PLANNING CAREER MOVES
WITH THE HELP OF THE INTERNET

There are numerous Web sites that offer one the opportunity to find out information about job openings and career opportunities in many professions
and occupations around the world. Whereas large organizations ranging
from governments to multinationals have job listings on their own Web
sites, smaller and medium-sized firms are usually trying to cross-list their
openings at one of the job boards.

<URL http://rescomp.standord.edu/jobs/>
   Job Hunt is a "metalist" for online sources about job openings and includes hotlinks to Europe and Asia, thereby providing one easy access to
   hundreds of online job boards and resource centers, including the best
   ones in your neck of the woods!!

Appendix H

Additional Sources
for Information

ELECTRONIC FORUMS

All the newsletters listed below are being moderated, that is, the editor checks what material is being included. The advantage of subscribing to such a moderated newsletter instead of an unmoderated Listserver is that most, if not all, insignificant and unrelated content is eliminated before the newsletter reaches your e-mailbox.

The best way to find out is to subscribe to three of the newsletters below and try them out for a few weeks. All newsletters are free!

Computer Underground Digest (CUD) is an open forum dedicated to sharing information among computerists and to the presentation and debate of diverse views. CUD material may be reprinted for nonprofit as long as the source is cited. Authors hold a presumptive copyright, and they should be contacted for reprint permission. It is assumed that nonpersonal mail to the moderators may be reprinted unless otherwise specified. Readers are encouraged to submit reasoned articles relating to computer culture and communication. Articles are preferred to short responses. Please avoid quoting previous posts unless absolutely necessary. Address:

To subscribe, send a one line message to:
LISTSERV@UIUCVMD.BITNET

or

LISTSERV@VMD.CSO.UIUC.EDU and in the body of the message type: SUB CUDIGEST Urs Gattiker
Frequency: Once a week, unless too much is happening

Cyberwire Dispatch provides important information about recent and still to come developments in the online industry with a nice twist of humor and inside information provided by Brock Meeks, a journalist.
To subscribe, send a message to:
Majordomo@cyberwerks.com and in the body of the message type: subscribe cwdl Urs Gattiker <Urs.Gattiker@unibwhamburg.de>
Frequency: Once a month, unless too much is happening

EDUCOM UPDATE is an electronic summary of organizational news and events on information technology and is provided as a service by Educom—a Washington, DC-based consortium of leading colleges and universities seeking to transform education through the use of information technology. The tidbits provided range from new product announcements to research findings of interest to executives around the world.
To subscribe, send a message to:
listproc@educom.edu and in the body of the message type: subscribe update FIRST NAME, LAST NAME
Frequency: Twice a month

EPIC is a nonprofit privacy advocate group in the United States. Its electronic newsletter Epic Alert appears about once every month and informs readers about new developments in cyberspace, on the Internet, and on electronic newsmedia. Address: Electronic Privacy Information Center (EPIC), Washington, DC, USA. E-Mail: info@epic.org
To subscribe, send a message to:
info@epic.org and in the body of the message type: SUB EPIC FIRSTNAME LASTNAME
Frequency: Once or twice a month

INFOSYS: The Electronic Newsletter for Information Systems provides important tidbits on a weekly basis about new developments around the world. It has reader contributions as well as short reviews of topic-related articles recently published in information systems-related publications and electronic newsletters around the world.
To subscribe, send a message to:

Infosys@American.edu and in the body of the message type: SUB
INFOSYS FIRSTNAME LASTNAME
Frequency: Once about every 10 days

Internet-Index is used for distributing the Internet Index, an occasional collection of Internet facts and statistics compiled by Win Treese (treese@OpenMarket.com). It is not a general discussion list.
To subscribe, send a message to:
Internetindexrequest@OpenMarket.com and in the body of the message type: SUBSCRIBE internet-index
Frequency: Once about every 30–60 days

Seidman's Online Insider is sponsored by NetGuide Magazine. It informs readers regularly about new developments in the online industry (e.g., what is CompuServe up to?). The information provided is very accurate and usually ahead of any other magazine or newspaper.
To subscribe to this newsletter, send an email message to:
LISTSERV@PEACH.EASE.LSOFT.COM and in the body of the message type: SUBSCRIBE ONLINE-L Urs Gattiker
Frequency: Once about every 2–6 weeks
A Web-based version of the newsletter is available at <URL http://www.netguidemag.com>.

SKYWRITING: The Internet Advertising Newsletter informs the reader about marketing and advertising issues as well as how to get the most out of one's Web and Internet activities. It was started in 1996 and is written by an advertising professional. Information and insights provided are very accurate.
To subscribe to this newsletter, send an email message to:
calumet@mindspring.comand with a subject header of SUBSCRIBE SKYWRITING and in the body of the message type:SUBSCRIBE SKYWRITING FIRSTNAME LASTNAME
Frequency: Once about every 3–4 weeks
You can see all back issues at <URL http://www.evspros. com/irc/skywrite/main.html>

The Computer Privacy Digest is a forum for discussion on the effect of technology on privacy or vice versa. The digest is moderated and gatewayed into the USENET newsgroup comp.society.privacy (Moderated). Address: Leonard P. Levine, Computer Privacy Digest, Professor of

Computer Science, University of WisconsinMilwaukee, Box 784, Milwaukee WI 53201, USA.

To subscribe, send a message to:

compprivacyrequest@uwm.edu and in the body of the message type:
SUBSCRIBE COMP-PRIVACY FIRSTNAME LASTNAME

Frequency: Two to four times a week; sometimes an issue is about one topic and has various contributions from readers around the world. Depending on the topic, some issues are of more interest than others but international viewpoints from readers are especially valuable.

TIM (Technology and Innovation Management) is the newsletter of the U.S. Academy of Management's TIM division. The newsletter has readers from around the world working and/or doing research in engineering, R&D, technology, and innovation management.

To subscribe, send a message to:

TIM@McMaster.CA and in the body of the message type: SUBSCRIBE TIM FIRST NAME LAST NAME

Frequency: About once every week

TIM-SecurityNews is a collaborative effort between TIM-Research and EICAR's Ad-hoc Working Group Trust in E-Commerce. WG Trust in E-Commerce of EICAR focuses on information safety, security, privacy, property as well as e-commerce issues and fighting spamming. TIM-Research is a virtual research organization focusing on matters involving new technologies and innovation.

If you would like to subscribe, please send an empty message to:

TIM-Security.Subscribe@News.WebUrb.dk

When you subscribe or unsubscribe, you will receive an e-mail message that asks you to confirm your request. This is simply done to prevent anybody else from signing you up or removing you from the list.

Using your Web browser you may also visit http://Security.WebUrb.dk (click on News) and look at some past issues of the newsletter (you need an Adobe Reader, which is free, to do so).

REFERENCES

Alba, R. D., Logan, J. R., & Bellair, P. E. (1994). Living with crime: The implications of racial/ethnic differences in suburban location. *Social Forces, 73*, 395–434.

American Psychological Society (1997). Reducing violence: A behavioral science research plan for violence. *APS Observer Special Issue, HCI Report 5*.

Andersen, E. S. (1994). *The evolution of credence goods: A transaction approach to product specification and quality control*. MAPP Working Paper #21, The Aarhus School of Business, Aarhus.

Angus, E., & McKay, D. (1994). *Canada's information highway*. Ottawa: New Media Branch and Information Technologies Industry Branch, Industry Canada.

Avrahami, R. (October 6. 1996). List sale case to the VA Supreme Court. *Computer Privacy Digest, 9* (23), 3–4.

Bauer, P. (November 21, 1995). Wettlauf ins All. Konzerne wetteifern mit Orbit (Competition into space. Organizations compete with satellite systems in orbit). *Die Welt*, 17.

Becker, H., & Fritzsche, D.J. (1987). A comparison of the ethical behaviour of American, French, and German managers. *Columbia Journal of World Business*, Winter, 87–95.

Bem, D. J. (1970). *Beliefs, attitudes and human affairs*. Belmont, CA: Brooks/Cole.

Beniger (1986). The control revolution: Technological and economic origins of the information society. Harvard University Press.

Bennett, C. J. (1992). *Regulating privacy*. Ithaca, NY: Cornell University Press.

Berger, P. (1997). *Redeeming laughter*. Berlin and New York: Walter de Gruyter.

Berger, P. L., & Luckmann, T. (1967). *The social construction of reality.* New York: Doubleday.

Bergmann, J. R. (1987). *Klatsch* (gossip). Berlin and New York: Walter de Gruyter.

Bhagat, R. S., & McQuaid, S. J. (1982). The role of subjective culture in organisations: A review and directions for future research. *Journal of Applied Psychology Monograph, 67,* 653–685.

Bikson, T.K., & Gutek, B.A. (1983). *Training in automated offices. An empirical study of design and methods,* Report No. WD–1904–R, Santa Monica, CA: Rand Cooperation.

Bill 68 1993, Chap. 17 (August 4, 1993). An act respecting the protection of personal information in the private sector. *Gazette Officielle du Québec,* 125, 4253–4279.

Blalock, H. M. (1984). *Basic dilemmas in the social sciences.* Beverly Hills: Sage Publications.

Blau, P. M. (1974). *On the nature of organizations.* New York: John Wiley and Sons.

Bowyer, K. (1995). Ethics and Computing: Living Responsibly in a Computerized World by Kevin Bowyer (Editor). New York, NY: IEEE Computer Society.

Brand, S. (1987). The politics of broadcatch. In *The media lab: Reinventing the future.* (pp. 201–223.) New York: Viking.

Brown, M. T. (1991). *Working ethics: Strategies for decision making and organizational responsibility.* San Francisco: Jossey-Bass Publishers.

Bundesministerium fuer Bildung, Wissenschaft, Forschung und Technologie (October 8, 1997). *Verordnung zur digitalen Signatur* (Signaturverordnung – SibV). (Regulation for digital signatures) (http://www.iid.de/rahmensigv.html) (February 28, 1998) (unofficial English translation can be found at http://ourworldcompuserve.com/homepages/ckuner).]

Bunn, M. (1994). Key aspects of organizational buying: Conceptualization and measurement. *Journal of the Academy of Marketing Science, 22,* 160–169.

Carey (1990). The language of technology: Talk, text, and template as metaphors for communication. In M. Medhurst, A. Gonzalez, & T. Peterson. (Eds.), *Communication and the culture of technology* (pp. 19–39). Pullman, WA: Washington State University Press.

Chang, T. Z. and Wildt, A. R. (1996). Impact of product information on the use of price as quality cue. *Psychology & Marketing, 13,* 55–75.

Chen, C., Lee, S. and Stevenson, H. W. (1995). Response style and cross-cultural comparisons of rating scales among East Asian and North American Students. *Psychological Science, 6,* 170–175.

Cleveland (1985). The twilight of hierarchy: Speculations on the global information society. *Public Administration Review,* Jan/Feb., 185–195.

Coffey, S., & Stipp H. (1997). The interactions between computer and television usage. *Journal of Advertising Research, 37*(2), 61–67.

Conger, S., Loch, K.D., Helft, B.L. (1995). Ethics and information technology use: A factor analysis of attitudes to computer use. *Information Systems Journal, 5*, 161–184.

Council of European Communities (November/December 1992, approved July 25, 1995). Amended proposal for a council directive (COM (92) 422 Final -SYN 283, October 15, 1992). *Transnational Data and Communications Report*, pp. 32–41. This proposal with slight modifications was approved by the Council of Ministers (July 25, 1995; European Commission Press Release: IP/95/822) and gave the 15 member states three years to implement the Directive.

Council of Ministers (August 5, 1997). *Schema di Regolamento Atti, documenti e contratti in forma elettronica* (regulatory framework for electronic documents and contracts in electronic form). Approved by the Italian Council of Ministers.

Council Regulation (EC) 3381/94, (December 19, 1994) *Setting up a community regime for the control of exports of dual-use goods*, OJ L 367/1, 31.12.94. Council Decision 94/942/CFSP, 19.12.94 establishes the lists of dual-use goods covered by the Regulation, OJ L 367/8, 31.12.94.

Cunningham, S. J. (1997). Interview with Jim Higgins on the Wired Wellington Project. *Technology Studies*, Vol. 4 158–163.

Denning, D. E., & Baugh, W. E. Jr. (1999). Hiding Crimes in Cyberspace, *Information, Communication and Society*, Vol. 2, No. 3, pp. 251–276.

Donaldson, T., & Preston, L. (1995). The Stakeholder Theory of the Corporation: Concepts, Evidence and Implications. *Academy of Management Review*, 20(1), pp. 65–91.

Donzé, L. (1993). The international Swiss telephone flows: The barriers to communication. *Communications*, 18, 291–305.

DTI (1997). Licensing of TTPs for the provision of encryption services - *DTI Public Consultation Paper #3*. Brussels: The Author (http://www.dti.gov.uk/pubs/) (December 17, 1997).

Dürrenberger, G., Jaeger, C., Bieri, L.,& Dahinden, U. (1995). Telework and vocational contact. *Technology Studies, 2*, 104–131.

Elizur, D., Borg, I., Hunt, R., & Beck, I. M. (1991). The structure of work values: A cross cultural comparison. *Journal of Organizational Behavior, 12*, 21–38.

England, G. W. (1987). Comparative patterns of work values among clerical employees in Oklahoma, the USA and other nations. *Oklahoma Business Bulletin*, 55(12), 17–22.

England, G. W., & Harpz, I. (1990). How working is defined: National contexts and demographic and organizational role influences. *Journal of Organizational Behavior, 11*, 253–266.

European Commission, *Telecommunications, Information Market and Exploitation of Research* (October 1997). Ensuring security and trust in electronic communication. Brussels: The Author (COM[97]503).

Fahs, R. (1997). Threats/vulnerabilities and risks. Proceedings of EICAR Security Workshop 1997 at Hamburg (pp. 8–25).

Feinberg, J. (1973). *Rights, justice, and the bounds of liberty: Essays in social philosophy.* Princeton, NJ: Princeton University Press.

Flew, A. (1971). *An introduction to Western philosophy: Ideas and argument from Plato to Sartre.* London, UK: Thames & Hudson.

Freeman, R. E. (1984). *Strategic Management: A Stakeholder Approach.* Boston, MA: Pitman.

Frigerio, C. G. (1995). Der neue Computerviren-Artikel im schweizerischen Strafgesetzbuch (the new computer virus article in the Swiss criminal code). *Proceedings of the 95 EICAR Conference - Improving Security of PCs and PC-Networks* (pp. T4-18 – T4-28, pp. 1–11).

Fulk, J. (1993). Social construction of communication technology. *Academy of Management Journal, 36*(5), 921–950.

Gattiker, U. E. (1990a). Where do we go from here? Directions for future research and managers. In: U. E. Gattiker & L. Larwood (eds.), *Studies in technological innovation and human resources* (Vol. 2) - *End-user training,* (pp. 287–303). Berlin & New York: Walter de Gruyter.

Gattiker, U. E. (ed.) (1990b). Individual differences and acquiring computer literacy: Are women more efficient than men? In: U. E. Gattiker, *Studies in technological innovation and human resources* (Vol. 2) - *End-user training,* (pp. 141–179). Berlin & New York: Walter de Gruyter.

Gattiker, U. E. (1990c). *Technology management in organizations.* Newbury Park, CA: Sage Publications.

Gattiker, U. E. (1992). Computer skill acquisition: Implications for end-user computing. *Journal of Management.*

Gattiker, U. E. (December, 1994). Im Cyberspace-Café. Bräuche und Missbräuche auf dem Information-Highway (The Cyberspace coffee-shop. Use and misuse on the information highway). *Tüte - Tübinger Termine (Special Edition on Daniel Foucault),* 46–50.

Gattiker, U. E. (1994). New technology, vocational training and recurrent education in Canada and Germany. In L. R. Gomez-Mejia & M. W. Lawless (Eds.), *Global high-technology management* (Vol. 4, Part B) (pp. 131–161). Greenwich, CT: JAI Press.

Gattiker, U. E. (1995). Firm and taxpayer returns from training of semiskilled employees. *Academy of Management Journal, 38,* 1152–1173.

Gattiker, U. E. (2000). *Internet challenges: Cultural, organizational and political issues.* Mahwah, NJ: Lawrence Erlbaum.

Gattiker, U. E., & Hedehus, D., with collaboration from members of EICAR Working Group 1 (Infosec: Information security and property rights). (1999). Virtual communities in Dream, Nightmare or Phantom for providers and users. In U. E. Gattiker, P. Pedersen & K. Petersen (Eds.), *Conference Proceedings EICAR International Conference,* ISBN: 87–987271–0–9.

Gattiker, U. E., & Hedehus, D. (1999). Managing virtual communities: Challenges and opportunities. In R. Berndt (Ed.), *Management strategies beyond 2000* (pp. 309–334). Berlin & New York: Springer-Verlag.

Gattiker, U. E., & Howg, L. W. (1990). Information technology and quality of worklife: Comparing users with non-users. *Journal of Business and Psychology*, 5, 237–260.

Gattiker, U. E., & Kelley, H. (1994). Techno-crime and terror against tomorrow's organisation: What about cyberpunks. In E. Raubold & K. Brunnstein (Eds), *Proceedings of the 13 World Computer Congress - IFIP Congress '94, Hamburg* (pp. 233–240). Amsterdam: Elsevier Science Publishers.

Gattiker, U. E., & Kelley, H. (1995). Morality and technology, or is it wrong to create and let loose a computer virus. In J. F. Nunamaker, Jr. & R. H. Sprague (Eds.), *Proceedings of the 28th Annual Hawaii International Conference on System Sciences 1995* (pp. 563–572). Las Alamitos, CA: IEEE Computer Society Press.

Gattiker, U. E., & Kelley, L. (1999). Morality and computers: Attitudes and differences in moral judgments across populations. *Information Systems Research*, 10, 450–475.

Gattiker, U. E., & Nelligan, T.W. (1988). Computerized offices in Canada and the United States: Investigating dispositional similarities and differences, *Journal of Organizational Behavior*, 9, 77–96.

Gattiker, U. E., & Willoughby, K. (1993). Technological competence, ethics, and the global village. Cross-national comparisons for organization research. In R. Golembiewski (Ed.), *Handbook of organizational behavior* (pp. 457–485). New York: Marcel Dekker.

Gattiker, U. E., Fahs, R., Blaha, J. and members of Working Group 1 of EICAR (1998). Medical data disclosure. Can encryption and effective public policy help in protecting individual privacy or is it a lost cause? *Proceedings of the 98 European Institute for Computer Anti-Virus Research (EICAR) Annual Conference - Improving Web and Internet Security* (pp. 79–93).

Gattiker, U. E., Gutek, B. A., & Berger, D. E. (1988). Office technology and employee attitudes. *Social Science Computer Review*, 6, 3, 327–339.

Gattiker, U. E., Janz, L., Kelley, H., & Schollmeyer, M. (1996). The Internet and Privacy: Do you know who's watching? *Business Quarterly*, 60(4), 1–6.

Gattiker, U. E., Greshake, J., Schwenteck, O., Janz, L., Holsten, H., & Kelb, J. (1997). Familiarity, knowledge and experience about privacy and security on the Internet: A cross-national study. Paper presented at the Academy of Management annual meeting, Boston.

Gattiker, U. E., Janz, L., Greshake, J, Kelb, J., Schwenteck, O., Holsten, H. (1996). Internet and Organisations: Social Aspects of Information Security. C. Schmid (Ed.), *Proceedings of the 96 European Institute for Computer Anti-Virus Research (EICAR) Annual Conference, Linz, Austria - Malicious software and the Internet* (pp. 185–203).

Gattiker, U. E., Schwenteck, O, Greshake, J., Janz, L., Kelb, J, & Holsten, H. (1997). The Internet community and ethics: A cross-national field study. *Proceedings of the Administrative Science Association of Canada - Technology and Innovation Management Division*, 12/Part 7, 77–86.

Gattiker, U. E., Kelb, J., Janz, L., Holsten, H., Greshake, J., Schwenteck, O., & Miller, J. (1997). Direct marketing and privacy for telephone and Internet users: A South African field study. *Global Business in Practice. Proceedings of the Tenth International Bled Electronic Commerce Conference, Bled, Slovenia*, 604–639.

Gauthier, D. (1986). *Morals by agreement*. Oxford, UK: Clarendon Press.

Gerster, J. (January 18, 1996). Auch die Kultur muß TV und Internet nutzen (Culture must also use TV and the Internet). *Die Welt*, p. 4.

Gewirth, A. (1978). *Reason and morality*. University of Chicago Press.

Gielen, U. P. (1982). A comparison of ideal self-ratings between American and German university students. In L. L. Adler (Ed.), *Cross-cultural research at issue* (pp. 275–288). New York: Academic Press.

Goldman, A. H. (1980). *The moral foundations of professional ethics*. Totowa, NJ: Rowman and Littlefield.

Griffeth, R. W., Hom, R. W., DeNisis, A. S., & Kirchner, W. K. (1985). A comparison of different methods of clustering countries on the basis of employee attitudes. *Human Relations, 38*, 813–840.

Grunert, K.G. (1997). What's in a steak? A cross-cultural study on the quality perception of beef. *Food Quality and Preference, 8* (3), 157–174.

Gutek, B. A., & Larwood, L. (1987). Information technology and working women in the USA. In M. J. Davidson and C. L. Cooper (Eds.), *Women and technology*, 71–94. Chichester, U.K.: John Wiley & Sons.

Haidt, J., Koller, S. H., & Dias, M. G. (1993). Affect culture, and morality, or is it wrong to eat your dog. *Journal of Personality and Social Psychology, 65*, 613–628.

Harper, L. (1993). Mind your own business. *Wall Street Journal*, October 5, 1993, p. A1.

Hersch, J. (1978). *Von der Einheit des Menschen* (From the unity of people). Zurich & Cologne: Benziger Verlag.

Hofstede, G. (1989). Organising for Cultural Diversity. *European Management Journal, 7*, 390–397.

Hofstede, G. (1991). *Cultures and Organisations. Intercultural Cooperation and Its Importance for Survival*. London: Harper Collins Publishers.

Holsten, H. M., Gattiker U. E., Janz, L., Kelb, J., Schwenteck, O., Greshake, J., & Miller, J. (1998). User attitudes toward governmental surveillance on the Internet. Paper presented at the Academy of Management annual meeting in San Diego.

Hulin, C. L., Drasgow, F., & Komocar, J. (1982). Applications of item response theory to analysis of attitude scale translations. *Journal of Applied Psychology, 67*, 818–825.

Igbarria, M., & Parasuraman, S. (1989). A path analytic study of individual characteristics, computer anxiety and attitudes toward microcomputers. *Journal of Management, 15*, 373–388.

Internet Domain Survey (July 1998). Internet domain survey, July 1998 [on-line]. Available: http://www.nw.com/zone/WWW/report.html

Internet users spending (March 4, 1997). Internet users spending less time watching TV. News – Morning Edition @ 8/27/97, Technology Section (via e-mail).

Iwata, E. (March 10, 1997). Internet commerce creates new legal issues. *NYT Computer News Daily* found on *Excite Live!* (Personal News Topics – Internet – March 12, 1997, http://www.excite.com).

Jacoby, J., Szybillo, G. J., & Busato-Schach, J. (1977). Information acquisition behavior in brand choice situations. *The Journal of Consumer Research, 3* (4), 209–216.

de Jasay, A. (1989). *Social contract, free ride: A study of the public goods problem.* Oxford, UK: Clarendon Press.

Jensen, J. (1990). Redeeming modernity: Contradictions in media criticism. Newbury Park, CA: Sage.

Johnson, D.G. (1989). The public–private status of transactions in computer networks. In C.C. Gould (Ed.), The information web: Ethical and social implications of computer networking, (pp. 37–55). Boulder, CO: Westview Press.

Johnson, B., & Rice, R. (1987). Managing organizational innovation: The evolution from word processing to office information systems. New York: Columbia University Press.

Kabay, M. (1998). Anonymity and pseudonymity in cyberspace: Deindividuation, incivility and lawlessness versus freedom and privacy. Proceedings of the 98 European Institute for Computer Anti-Virus Research (EICAR) Annual Conference Improving Web and Internet Security (pp. 1–20).

Kahneman, D., & Tversky, A. (1979). Prospect theory: An analysis of decision under risk. *Econometrica, 47,* 263–291.

Kallman, E. A. (1992). Developing a code for ethical computer use. *Journal of Systems Software, 17,* 69–74.

Karnow, C. E. A. (July 1994). *Recombinant culture: Crime in the digital network.* Paper presented at Defcon II, Las Vegas.

Keen, P. (1991). Managing the economics of information capital. Chapter 6, *Shaping the future: Business design through informaiton technology.* (Pp. 141–178.) Boston, MA: Harvard Business School Press.

Kelley, H., Gattiker, U. E., Paulson, D., & Bathnagar, D. (1994). End-user attitudes and information systems: A cross-national study. Paper presented at the Annual Meeting of the Administrative Sciences Association of Canada, Halifax.

Kjaerulf, J. (1998). Scope of the study, motivation and prior research in the field. Proposal for Ph.D. study, July 1998.

Kohlberg, I. (1969). Stage and sequence: The cognitive–development approach to socialization. In D. A. Goslin (Ed.), Handbook of moralization theory and research (pp. 347–480). Chicago, IL: Rand McNally.

Laurin, F., & Froste, C. (1997). Secret Swedish e-mail can be read by the USA. *Svenska Dagbladet*, November 18, 1997, (ukcrypto@maillist.ox.ac.uk, December 26, 1997).

Lawson, R. (1997). Consumer decision making within a goal-driven framework. *Psychology & Marketing, 14*, 427–449.

Logan, R., Snarey, J., & Schrader, D. (1990). Autonomous versus heterogenous moral judgement types. A longitudinal cross-cultural study. *Journal of Cross-Cultural Psychology, 21*, 71–89.

Loi, N. 90–1170 (29 December 1997). Article No. 28. Telecommunications law. (http://www.legifrance.fr). (http://www.telecom.gouv.fr/ francais/activ/ telecom /nloi17.htm) (December 17, 1997).

Lyotard, J-F. (1984). *The postmodern condition: A report on knowledge* (translation from French by Geoff Bennington and Brian Massumi, Foreword by Frederic Jameson). Minneapolis: University of Minnesota Press.

Lyotard, J.-F. (1988). Unpublished conversations with René Guiffrey (translated from French by Geoff Bennington). In: G. Bennington (ed.), *Lyotard: Writing the event*. Manchester, UK: Manchester University Press.

van Maanen, J., & Barley, S. R. (1985). Cultural Organization: Fragments of a Theory. In Frost, P.J., Moore, L.F. Louis, M.R., Lundberg, C. C. and Martin, J. (eds.). *Organizational Culture*, 31–55. Newbury Park: Sage Publications.

March, J. G., & Simon, H. A. (1958). *Organizations*. New York: John Wiley.

Matsueda, R. L., & Heimer, K. (1987). Race, family structure, and delinquency: A test of differential association and social control theories, *American Sociological Review, 52*, 826–840.

Mayhew, B. H., & Levinger, R.L. (1976). Size and the density of interacting in human aggregates. *American Journal of Sociology, 82*, 86–110.

McCarty, J. A., & Shrum, L. J. (1993). The role of personal values and demographics in predicting television viewing behaviour: Implications for theory and application. *Journal of Advertising, 22* (4), 77–101.

McClosky, H., & Brill, A. (1983). Dimensions of tolerance: What Americans believe about civil liberties. New York: Russel Sage Foundation.

McDonald, S. (1997). The once and future Web: scenarios for advertisers. *Journal of Advertising Research, 37* (2), 21–28.

Meeks, B. (January 1997). Jacking in from the "Devil's own dialtone" port: *CyberWire Dispatch*, 1–2.

Meyrowitz, J. (1985). *No sense of place: The impact of electronic media on social behavior*. New York: Oxford University Press.

Miller, J. G., Bersoff, D. M., & Harwood, R. L. (1990). Perceptions of social responsibilities in India and the United States: Moral imperatives or personal decisions? Journal of Personality and Social Psychology, 58, 33–47.

Mitchell, R. K., Agle, B. R., & Wood, D. J. (1997). Toward a theory of stakeholder identification and salience: Defining the principle of who and what really counts. *Academy of Management Review*, 22, 853–886.

Molander, E. A. (1987). A paradigm for design, promulgation and enforcement of ethical codes. *Journal of Business Ethics, 6*, 619–631.

Mulgan, G. (1990). Communication and control: Networks and the new economics of communication. New York: Guilford Press.

Munson, W. (1993). *All talk*. Philadelphia: Temple University Press.

Nelson, P. (1970). Information and consumer behaviour. *Journal of Political Economy, 78*, 311–329.

Newton, B. J., & Buck, E. B. (1985). Television as significant other. Its relationship to self-descriptors in five countries. *Journal of Cross-Cultural Psychology, 16*, 289–312.

Nino, C. S. (1991). The ethics of human rights. London, UK and New York: Clarendon Press.

Nissen, T. (1998). Chrysler's Eaton to dealers don't resist Internet. *Infobeat News – Morning Edition*, February 2, 1998; <http://www.infobeat.com/stories/cgi/story.cgi?id=2552675071–e4c <Accessed February 2 , 1998>.

Nozick, R. (1974). *Anarchy, state, and utopia*. New York: Basic Books.

NZZ. (Jan. 9, 1988). Probleme der flexiblen Arbeit in den Niederlanden (Problems with flexible work in the Netherlands, p. 20.

Nucci, L. (1981). Conception of personal issues: A domain distinct from moral or societal concepts. *Child Development, 52*, 114–121.

OECD (1997). Report on background and issues of cryptography policy. Paris: The Author.

Okoshen, H. R. (1996). A cross cultural comparison of ethical perspectives and decision approaches of business students: United States of America versus New Zealand, *Journal of Business Ethics, 15*, 537–549.

Organizations push for worldwide net guidelines (March 8, 1997). *Excite Live!* (Technology — March 8, 1997, http://www.excite.com).

Oskamp, S. (1997). *Attitudes and opinions*. Englewood Cliffs, NJ: Prentice Hall.

Piaget, J. (1965). *The moral judgment of the child* (Translated from French to English by Marjorie Gabain). New York: Free Press.

Pierce, M. A., & Henry, J. W. (1996). Computer ethics: The role of personal, informal, and formal codes. *Journal of Business Ethics, 15*, 425–437.

Plomin, R., & Rende, R. (1991). Human behavioral genetics. Annual Review of Psychology, 42, 161–182.

Rakow, L. F., & Navarro, V. (1993). Remote mothering and the parallel shift: Women meet the cellular telephone. *Critical Studies in Mass Communication, 10*, 144–157.

Ramsy, I. (1985). Framework for regulation of the consumer marketplace. *Journal of Consumer Policy*, 8, 353–372.

Ravi, D. (1997). Consumer preference for a no-choice option. *Journal of Consumer Research, 24*, 215–231.

Rawls, J. (1971). A Theory of Justice. Cambridge, MA: Harvard University Press.

Rest, J. R., Thoma, S. J., Moon, Y. L., & Getz, I. (1986). Different cultures, sexes, and religions. In J.R. Rest (Ed.), *Moral development. Advances in research and theory*, (pp. 89–132). New York: Praeger Publishers.

Rhodes, S. R. (1983). Age related differences in work attitudes and behaviour: A review and conceptual analysis. *Psychological Bulletin, 93*, 328–367.

Rice, R. E. (1987). Computer-mediated communication systems and organizational innovation. *Journal of Communication, 37*, (4), 65–94.

Rice, R. E. (1992). Contexts of research on organizational computer-mediated communication. (Pp. 113–144). United Kingdom: Harvester-Wheatsheaf.

Rice, R., & Gattiker, U. E. (2000). Computer-mediated organizational communication and structure. In F. Jablin & L. Putnam (Eds.), *New handbook of organization communication*. Newbury Park: Sage Publications.

Rice, R. E., & Steinfield, C. (1994). New forms of organizational communication via electronic mail and voice messaging. In J. Erik Andriessen & R. Roe (eds.) *Telematics and work*. (Pp. 109–137.) NJ: Lawrence Erlbaum.

Rodgers, R. (1992). Antidotes for the idiot's paradox. In U. E. Gattiker (ed.), *Studies in technological innovation and human resources* (Vol. 3) - *Technology-mediated communication*, (pp. 227–271). Berlin & New York: Walter de Gruyter.

Rogers, E. M., 1962. *Diffusion of innovations*. New York: The Free Press.

Ronen, S., & Shenkar, O. (1985). Clustering countries on attitudinal dimensions: A review and synthesis. *Academy of Management Review, 10*, 435–454.

Rokeach, M. (1980). *Beliefs, attitudes, and values*. San Francisco: Jossey-Bass.

Rosen, D. L., & Olshavsky, R. W. (1987). The dual role of informational social influence: Implications for marketing management. *Journal of Business Research, 15*, 123–144.

Sadalla, E. K. (1978). Population size, structural differentiation, and human behaviour. *Environment and Behaviour, 10*, 271–291.

Sarlo, C. A. (1992). *Poverty in Canada*. Vancouver: The Fraser Institute.

Schlossberg, H. (1993). Victims tired of researchers getting away with murder. *Marketing News*, August 16, A16, 1.

Schmitt, B. H., & Shultz, C. J. II (1995). Situational effects on brand preferences for image products. *Psychology & Marketing, 12*, 433–446.

DeSerpa, A. C. (1994). Pigou & Coase. A mathematical reconciliation. *Journal of Public Economics, 54*, 267–286.

Shweder, R.A., Mahapatra, M., & Miller, J.G. (1987). Culture and moral development. In J. Kagan & S. Lamb (Eds.), *The emergence of morality in young children* (1–90). Chicago: University of Chicago Press.

Silverman, D. (1970). *The theory of organizations*. London: Heinemann.

Sim, S. (1992). Beyond aesthetics. Confrontations with poststructuralism and postmodernism. Toronto: University of Toronto Press.

Sitkin, S. B., & Weingart, L. R. (1995). Determinants of risky decision-making behaviour: A test of the mediating role of risk perceptions and propensity. *Academy of Management Journal, 38*, 1573–1592.

Smith, B. H. (1991). Note. Anxiety as a cost of commuting to work. *Journal of Urban Economics*, 29, 260–266.

Smith, G. E. (1996). Framing in advertising and the moderating impact of consumer education. *Journal of Advertising Research, 36* (5), 49–64.

Soe, L., & Marcus, M. L. (1993). Technological or social utility? Unraveling explanations of e-mail, emai,l and fax use. *The Information Society*, 9, 213–236.

Sproull, S. & Kiesler, S. (1991) *Connections: New ways of working in the networked organization.* Cambridge, MA: MIT Press.

Stead, W. E., & Stead, J. G. (1992). Management for a small planet. Newbury Park, CA: Sage Publications.

Stone, E. F., Gardner, D. G., Gueutal, H. G., & McClure, S. A. (1983). A field experiment comparing information– privacy values, beliefs, and attitudes across several types of organisations. *Journal of Applied Psychology, 68*, 3, 459–468.

Stone, E. F., & Stone, D. L. (1990). Privacy in organisations: Theoretical issues, research findings, and protection mechanisms. *Research in Personnel and Human Resources Management, 8*, 349–411.

Strack, F., & Förster, J. (1995). Reporting recollective experiences: Direct access to memory systems? Psychological Science, 6, 352–358.

van Swaay, M. (1995). The Value and Protection of Privacy. *Computer Networks and ISDN Systems, 26* (Suppl. 4), 149–155.

Taylor, P. (December 4, 1995). Oracle to put every UK school on the Internet. *Financial Times*, pp. 7, 13.

Thorel, J. (18 December, 1997). Frenchy Cryptosoap #123578. *Lambda*, 3.08, p. 1 (see also http://www.freenix.fr/netizen/chiffre/avis-cssp.html).

Thumin, F. J., Johnson, J. H., Kuehl, C. , & Jiang, W. (1995). Corporate values as related to occupation, gender, age and company size. *Journal of Psychology, 129*, 389–400.

Treese, W. (January 26, 1998). Percentage of purchasing. *The Internet Index*, No. 21, p. 1.

Trevino, L., & Webster, J. (1992). Flow in computer-mediated communication: Electronic mail and voice mail evaluation and impacts. *Communication Research*, 19, 539–573.

Triandis, H. C. (1977). Cross-cultural social and personality psychology. *Personality and Social Psychology Bulletin*, 56, 143–158.

Triandis, H. C., & Vassiliou, V. (1972). Interpersonal influence and employee selection in two cultures. *Journal of Applied Psychology, 56*, 140–145.

Turiel, E. (1983). The development of social knowledge: Morality and convention. Cambridge, UK: Cambridge University Press.

Turiel, E., Killen, M. & Helwig, C. C. (1987). Morality: Its structure, functions, and vagaries. In: J. Kaan & S. Lamb (eds.), *The emergence of morality in young children*, (pp. 155–243).

Ulhøi, J. P. (1997). A stakeholder approach to green innovation. In Proceedings of The Fourth International Meeting of the Decision Science Institute, Sydney.

Ulhøi, J. P., & Madsen, H. (1998). Greening of industry in a push–pull stakeholder perspective. Working paper, The Aarhus School of Business.

Venkataramani, J. G., Kamel, G., & Jacoby, J. (1997). A varying-parameter averaging model of on-line brand evaluations. *Journal of Consumer Research, 24*, 232–247.

Vicini, J. (March 16, 1997). High Court to hear Internet free-speech case. *Reuter* found on *Excite Live!* (Internet – March 17, http://www.excite.com)

Webster's College Dictionary (1991). New York: Random House.

Weisband, S. P., & Reining, B. A. (1995). Managing user perceptions of E-mail privacy. Communications of the ACM, 38, 12, 40–47.

Wijkander, H. (1985). Correcting externalities through taxes on/subsidies to related goods. *Journal of public Economics, 28*, 111–125.

Wilde, L. L. (1980). The economics of consumer information acquisition. *Journal of Business, 53*, 143–158.

Wildeman, H. (1997). Fertigungsstrategien: Reorganisationskonzepte fuer eine schlanke Production und Zulieferung (Production strategies: Concepts for re-organizing to achieve a lean production and supply chain).

Wilder, G., Mackie, D., & Cooper, J. (1985). Gender and computer: Two surveys of computer-related attitudes. *Sex Roles, 13*, 215–238.

Willoughby, K. W. (1990). *Technology choice.* Boulder, CO: Westview Press.

Wilson, T. C. (1995). Urbanism and unconventionality: The case of sexual behaviour. *Social Science Quarterly, 76*, 346–363.

Wirl, F. (1994). Pigouvian taxation of energy for flow and stock externalities and strategic, noncompetitive energy pricing. *Journal of Environmental Economics and Management, 26*, 1–18.

Yates, J., & Benjamin, R. (1991). The past and present as a window on the future. In M. S. Scott Morton (ed.). The corporation of the 1990s: Information technology and organizational transformation. (Pp. 61–92.) NY: Oxford U. Press.

Yates J., & Orlikowski, W. J. (1992). Generes of organizational communication: A structural approach to studying communication and media. Academy of Management Review, 17, 299–326.

Zedeck, S., & Cascio, W. F. (1984). Psychological issues in personnel decisions. *Annual Review of Psychology, 35*, 461–518.

Geographical Places/Locations Mentioned in Text

Aalborg, 37, 38, 207, 208, 234
Aarhus, 207
Africa, 32, 40
African, 32
Albania, 93
Alberta, 38, 183, 222, 233
America, 4, 62, 87
American, 68, 84, 91, 92, 107, 130, 131,
        134, 135, 137, 147, 154, 155,
        176, 192
Argentinian, 37
Asia, 155, 238
Asian, 55
Atlantic, 95
Australia, 65, 180
Austrian, 136
Bangalore, 66
Belgium, 51
Bologna, 15
Brazil, 63
British, 33, 84, 92, 222, 223
California, 168, 183, 213, 214, 233
Canada, 38, 41, 47, 48, 60, 63, 65, 72, 91,
        95, 116, 120, 131, 138, 160, 165,
        191, 200, 207, 222, 233
Canadian, 32, 43, 47, 55, 60, 61, 62, 68,
        131, 152, 165, 186, 228, 233
Cape Town, 40
China, 109, 189

Chinese, 93, 95, 109, 175, 189
Copenhagen, 15, 168
Danish, 34, 89, 180
Delaware, 61
Denmark, 23, 34, 38, 39, 51, 58, 59, 68,
        92, 97, 180, 181, 183, 186, 207,
        208, 234
Detroit, 12
Dutch, 51, 97
Dutch Guyana, 23
England, 10, 31, 44, 47, 53, 58, 68, 180,
        189, 222, 223
English, 23, 24, 51, 58, 68, 84, 91, 93, 94,
        95, 149, 152, 175, 181, 189, 191,
        193
Europe, 34, 37, 39, 50, 63, 66, 68, 70, 84,
        93, 94, 97, 101, 149, 155, 180,
        238
European, 1, 4, 10, 31, 45, 46, 47, 48, 49,
        50, 51, 52, 55, 59, 65, 66, 67, 69,
        81, 84, 85, 89, 95, 97, 116, 117,
        118, 129, 170, 206, 228, 231,
        232, 233, 235
Finland, 34, 180
France, 51, 52, 54, 55, 116
Frankfurt, 55
French, 51, 55, 93, 95, 109, 116, 118, 152,
        175
Geneva, 118, 206

255

Organizations and Associations Mentioned in Text

Author Index

259

Subject Index